"This book is a must-read for al
features of patristic exegesis. A...... p.o......s a j.......s and much-needed
defense against making the early Fathers conform to various conservative
versions of interpreting Scripture. Using the Genesis creation account, the
reader is invited to see that the ancients were far more imaginative and
biblically minded than we credit them."

D. H. Williams, professor of patristics and historical theology at Baylor
University

"This is a brave and much-needed book. A church that tries to ignore the
Fathers of the Church deprives itself of a valuable resource. Craig Allert
seeks to show how attention to how the Fathers understood Genesis 1
deepens our own understanding of creation. He cuts through a lot of misunderstanding and ignorance of the Fathers enabling us to hear them once
more. Professor Allert's proposal is not so much 'Back to the Fathers' as
'Forward with the Fathers.'"

Andrew Louth, professor emeritus at Durham University

"Navigating the first chapter of Genesis, especially in light of present-day
controversies, is tricky business. Craig Allert's presentation of early
Christian readings of this text will help readers to understand ancient perspectives and their applicability to present concerns."

Christopher A. Hall, distinguished professor of theology emeritus at Eastern
University, and president of Renovaré USA

"In this book Allert explains the concept of church father and their interpretations of Genesis, discusses the erroneous uses of the Fathers, and directs us to a fuller understanding of them. He treats Basil and Augustine in
particular, especially Basil's homilies on Genesis. This book is an excellent
antidote to fundamentalist and creationist misreadings of Genesis 1."

Mark Sheridan, professor and rector emeritus, Pontificio Ateneo S. Anselmo,
Rome, Italy

EARLY CHRISTIAN
READINGS
of

GENESIS
ONE

Patristic

Exegesis and

Literal Interpretation

CRAIG D. ALLERT

An imprint of InterVarsity Press
Downers Grove, Illinois

InterVarsity Press
P.O. Box 1400, Downers Grove, IL 60515-1426
ivpress.com
email@ivpress.com

InterVarsity Press® is the book-publishing division of InterVarsity Christian Fellowship/USA®, a movement of
students and faculty active on campus at hundreds of universities, colleges, and schools of nursing in the United
States of America, and a member movement of the International Fellowship of Evangelical Students. For
information about local and regional activities, visit intervarsity.org.

Cover design: David Fassett
Interior design: Daniel van Loon
Images: Latin manuscript: © DEA / G. DAGLI ORTI / Getty Images
     Greek fresco: © jovanjaric / iStock / Getty Images Plus
     solar system: © adventtr / E+ / Getty Images
     paper texture: © ke77kz / iStock / Getty Images Plus

ISBN 978-0-8308-5201-7 (print)
ISBN 978-0-8308-8783-5 (digital)

Printed in the United States of America ∞

InterVarsity Press is committed to ecological stewardship and to the conservation of natural resources in all our
operations. This book was printed using sustainably sourced paper.

Library of Congress Cataloging-in-Publication Data

Names: Allert, Craig D., author.
Title: Early Christian readings of Genesis one : patristic exegesis and
  literal interpretation / Craig D. Allert.
Description: Downers Grove : InterVarsity Press, 2018. | Includes
  bibliographical references and index.
Identifiers: LCCN 2018012251 (print) | LCCN 2018019655 (ebook) | ISBN
  9780830887835 (eBook) | ISBN 9780830852017 (pbk. : alk. paper)
Subjects: LCSH: Bible. Genesis, I — Criticism, interpretation,
  etc. — History — Early church, ca. 30-600.
Classification: LCC BS1235.52 (ebook) | LCC BS1235.52 .A45 2018 (print) | DDC
  222/.110609015 — dc23
LC record available at https://lccn.loc.gov/2018012251

P	22	21	20	19	18	17	16	15	14	13	12	11	10	9	8	7	6	5	4	3	2	1
Y	37	36	35	34	33	32	31	30	29	28	27	26	25	24	23	22	21	20	19	18		

To Dr. Peter Flint (1951–2016)

CONTENTS

ACKNOWLEDGMENTS

There is a reason some things are said often. In the case of publishing a book, it is often stated that it could not have become a reality without the help, support, and patience of others. This book is no different. Foundationally, it was made possible through the support of a grant from the BioLogos Foundation's Evolution and Christian Faith program, which allowed me to lighten a heavy teaching and administrative load so that I could better concentrate on this project. For this I am very grateful.

There are several people at Trinity Western University to whom I owe much thanks. I thank Provost Dr. Bob Wood and my former dean, Dr. Bob Burkinshaw, for their support of the project and for working to make course and administrative relief possible. My new (but now former) dean, Dr. Myron Penner, continued that support. Thanks also go to Eve Stringham and Sue Funk in the Office of Research for administering the details of the grant so I could focus on the research. Sue provided attentive and very helpful advice and guidance in these matters, for which I am very grateful. To my colleagues in the Religious Studies Department — I am so thankful for all of you. It is a blessing to teach and research in such a vibrant, committed, and collegial department.

InterVarsity Press's Dan Reid offered some very timely and helpful suggestions after he read a draft of the manuscript. The suggestions both saved me some embarrassment and have made the book much better.

I thank many close friends who have consistently taken an interest in this project. Their questions about the progress and the general topic grew out of an excitement they had for it and served as a source of encouragement during times when I wondered if the book would ever see the light of day.

My wife, Corinne, and our sons, Ty and Zach, deserve a huge thank you. There were times when this project took me away from certain family necessities. In particular, Corinne often recognized the need for extended periods of time of devotion to the book and gave me the freedom to spend those hours confident in her support.

As I write these words, the Religious Studies Department at Trinity Western University is at the one-year anniversary of the sudden death of our esteemed colleague Dr. Peter Flint, who passed away on November 3, 2016. Although Peter and I were in different academic areas, he was always a strong source of encouragement to me personally and academically. He seldom let a passing hello or longer conversation go by without saying an encouraging word about my research or the job I was doing in my administrative duties for the department. Peter will be greatly missed, and it is to him that I dedicate this book.

ABBREVIATIONS

AiG   Answers in Genesis

ANF   Alexander Roberts and James Donaldson, eds. Ante-Nicene Fathers. 10 vols. Buffalo, NY: Christian Literature, 1885–1887; reprint, Grand Rapids: Eerdmans, 1951–1956; reprint, Peabody, MA: Hendrickson, 1994.

CMI   Creation Ministries International

FC   Fathers of the Church: A New Translation. Washington, DC: Catholic University of America Press, 1946–.

GH   grammatical-historical

HC   historical criticism

ICR   Institute for Creation Research

LXX   Septuagint

NPNF[1]   Philip Schaff et al. A Select Library of the Nicene and Post-Nicene Fathers of the Christian Church. 1st series. 14 vols. Buffalo, NY: Christian Literature, 1887–1894; reprint, Grand Rapids: Eerdmans, 1952–1956; reprint, Peabody, MA: Hendrickson, 1994.

NPNF[2]   Philip Schaff et al. A Select Library of the Nicene and Post-Nicene Fathers of the Christian Church. 2nd series. 14 vols. Buffalo, NY: Christian Literature, 1887–1894; reprint, Grand Rapids: Eerdmans, 1952–1956; reprint, Peabody, MA: Hendrickson, 1994.

OECT   Oxford Early Christian Texts. Oxford: Oxford University Press, 1970–.

PPS   Popular Patristics Series. Crestwood, NY: St. Vladimir's Seminary Press, 1977–.

TWSA   The Works of Saint Augustine: A Translation for the 21st Century. Hyde Park, NY: New City Press, 1990–.

INTRODUCTION

A nyone who has had the misfortune of being misunderstood or having their words used in unintended ways will know the helplessness and frustration that can result. As a professor of theology at an evangelical Christian liberal arts university in Canada, I know firsthand what this is like. Of the many times I have experienced this feeling, one example following the publication of my book *A High View of Scripture?* stands out.[1] Even though I received encouraging words and reviews from many of my evangelical brothers and sisters, the book also generated a fair amount of dissatisfaction and angst over some of my conclusions.

One negative reaction came from an American pastor in the form of an unpublished book review. Of the several complaints he had with the book, one stood out because of its misleading nature. Near the end of his review he accused me of caring more about academic respectability than the good of the church. In his words:

> One of the driving forces behind how he [Allert] sees the task of theology and scholarship is respectability in the larger academic community. Speaking of conservative views of scripture he comments: "If one unreservedly throws

---

[1]Craig D. Allert, *A High View of Scripture? The Authority of the Bible and the Formation of the New Testament Canon*, Evangelical *Ressourcement* Series (Grand Rapids: Baker Academic, 2007).

in one's lot with the right, *severe criticisms would center on that individual con-cerning the claims of biblical scholarship.*" (emphasis added)

This is a classic example of misleading readers because of what the author does *not* include from the source. While it is true that the pastor cites the words as written, he does not indicate, as I do in the book, that I am summarizing the words of the well-known evangelical scholar I. Howard Marshall. I was relating to my readers the tension expressed by Marshall in his book on biblical inspiration as he set out to explain this doctrine to an evangelical audience.[2] In addition to this lack of recognition, the pastor also conveniently excluded much of the quotation that balances the point I was making with reference to Marshall and inspiration. The full quotation from my book, with the section quoted by the pastor italicized, is as follows:

> I am very aware of the fact that entering into a presentation and discussion of this sort can be a risky venture. This is so not only because of the divergent views prevalent within the topic, but also because of the dedication and dogma with which certain views are held. The comment of one well-known New Testament scholar comes to mind as I embark on this endeavor. I. H. Marshall believes that anyone daring enough to express his views on the nature of the Bible has a particularly difficult path to tread.[3] His path does not merely run along the edge of a cliff, with the sea below on one side; he is traversing an edge with the possibility of going astray both to the right and to the left. If one deviates to the left and suggests that not all of what Scripture says is true in the strictest sense of the term, that individual will come under severe criticism from the right, not simply for saying so, but also for saying so as a confessed evangelical. *If one unreservedly throws in one's lot with the right, severe criticisms would center on that individual concerning the claim of biblical scholarship.*[4]

I must admit feeling angry at this accusation that so blatantly twisted the context and intent behind these words. To accuse me of caring

---

[2]I. Howard Marshall, *Biblical Inspiration*, Biblical and Theological Classics Library (Carlisle, UK: Paternoster, 1995), 7.
[3]Marshall, *Biblical Inspiration*, 7.
[4]Allert, *High View of Scripture?*, 14-15.

more about academic respectability than anything else by using a fragment from an extended quotation is, to be frank, careless, irresponsible, and dishonest. It is a move intended to sway the audience to the author's point of view by not letting the details of context get in the way of the point he was intent on making.

Unfortunately, irresponsible use of sources is not uncommon in our evangelical world. I do not know the reasons why some choose to take this line of "argument," and it is not this book's intent to investigate it, but it certainly needs to be addressed in the specific instances it occurs. The problem of irresponsible use of sources intersects with the purpose of this book. Broadly stated, my intention is to give a window into the strange new world of the church fathers and how they understood creation themes in Genesis 1. I say "strange new world" because for many evangelicals it is just that. Most have heard of the church fathers, and some have heard the names Augustine or Origen. In rare cases appeal is even made to them in support of an argument. But this is usually as far as it goes.

This lack of knowledge is a problem when trusted sources are irresponsible in how they appeal to the church fathers. This is precisely what is happening in some creation science appropriations of patristic sources. Rather than using the deliberate and patient approach necessary to this strange new world, creation science advocates decontextualize and proof-text the early leaders of the church to show that they read Scripture and understood Genesis 1 no differently than they themselves do today. Rather than trying understand the church fathers on their own terms, as pastors and churchmen, they are treated as ammunition in a battle they knew nothing about. To treat the Fathers so irresponsibly is a disservice both to them and the church they served. Further, it fails to accurately discern how we may grasp our own debt to them and properly appropriate them to inform our contemporary Christianity.

In some ways, the irresponsible appropriation of which I write does not surprise me. We evangelicals do not often realize how culture affects our own understanding of the Bible. We have the notion that we can simply read and understand Scripture clearly without any preconceived ideas or influences. Disagreements with our interpretation, in this way of thinking, must be the result of *others* bringing preconceived ideas or influences to their reading – certainly not our own! This attitude is just as alive in religion and science as it is in every other arena. This is why evangelical historian Mark Noll sought to identify attitudes, assumptions, and convictions that influence North American Protestants' interaction with science.[5] Doing this, he states, helps us to "realize how much pre-commitments affect contested issues of science and religion and to urge as much self-critical self-consciousness as possible when approaching such questions."[6] It will not do to claim a bird's-eye view, where we are uninfluenced by the particularity of our own historical locatedness over against others who disagree with us. A better approach would be to understand our influences and assumptions as best we can, just as we understand the influences and assumptions of those we use as sources.

As I prepared this book I was struck, and sometimes even appalled, at the way the church fathers were used as ammunition by some in the creation/evolution debate. Many examples are detailed in chapter two, "How *Not* to Read the Fathers." But before we can get there it is necessary to understand who the church fathers were and why they remain important for Christianity today, which is explained in the opening chapter, "Who Are the Church Fathers, and Why Should I Care?"

Foundational to the creation/evolution debate is the interpretation of Scripture. Assumptions are made and often transferred to early biblical interpreters. Descriptions like "literal," "plain/common meaning," and "allegory" are thrown around as if they are understood in the same way across Christian history. When this occurs, the risk

---

[5]Mark A. Noll, "Evangelicals, Creation, and Scripture: Legacies from a Long History," *Perspectives on Science and Christian Faith* 63, no. 3 (2011): 147-58.
[6]Ibid., 147.

of significant misunderstanding is very high. In order to address the misunderstandings and misrepresentations of biblical interpretation in the age of the Fathers, chapter three asks, "What does 'literal' mean?" Some who want to go immediately to the nitty gritty of the creation/evolution issue may be frustrated with this chapter because it does not deal directly with the debate. But I beg the reader's patience because it is here that foundational interpretational issues and definitions are addressed and clarified, thereby giving us the context within which to understand the Fathers' approach to Scripture that is too often glossed.

Chapter four, "Basil the Literalist?" builds on this foundation by addressing a specific example in Basil of Caesarea. As we will learn, Basil is often paraded as a champion of literal interpretation in the modern sense by advocates of a creation science approach to Genesis 1. In his homilies on the six days of creation, Basil has some remarks about allegorical interpretation that are seen by some as an outright rejection of it in favor of the literal. We will see that, over against simple proof-texting, proper context is necessary to understand that his words were neither an acceptance of literal interpretation in our modern sense nor an outright rejection of allegorical interpretation.

I shift the approach in the next three chapters to organize them more around themes or topics of importance in the Fathers' understanding of Genesis 1–2. In all three chapters my intent is to allow the reader to see the flow and context of the arguments as they are expressed around these themes. In other words, seeing how some Fathers approached and discussed issues that have become important in the contemporary creation/evolution debate may allow us to learn from another Christian perspective. The chapters are deliberate and follow the flow of argument in the limited group of Fathers I have chosen. This allows us to see context – what they emphasized and how they did so. Chapter five, "Creation out of Nothing," thus examines the background and some principal issues regarding the understanding of creation in

Theophilus, Ephrem, and Basil. Chapter six, "The Days of Genesis," surveys the same three Fathers concerning the topic of the days. Chapter seven looks at how Augustine discussed the first words of Genesis, "In the beginning."

The final chapter, "On Being like Moses," seeks to show the reader how one Father (Basil of Caesarea) uses the creation-of-humanity narrative in Genesis 1 and 2 as a foundational anthropology that has implications for interpreting Scripture. It is meant as a small offering to reconsider how the biblical creation account can function in the life of a Christian by looking at what Basil emphasizes in his interpretation of the creation of humanity.

This book does not pretend to be exhaustive. It is a small step and humble call to a responsible appropriation of our Christian past. The small sampling of church fathers examined in this book should demonstrate that they cannot simply be mined for proofs showing how they agree with a particular side in the creation/evolution debate. This practice flattens their importance to contemporary Christianity and makes our modern concerns their concerns. It does not allow them to speak to us about what they believed important for the life of the church and reduces them to ammunition in a debate that is peripheral to the well-being of the church.

As we will see, a foundational message of the church fathers studied in this book is that God is Creator, and his creation points us back to himself. Further, our place in that creation is one that requires restoration and reconciliation to him. This should serve to communicate something about the relative importance of creation in the life of the Christian. The Fathers can help us understand this if we approach them responsibly. What does a responsible appropriation of the Fathers look like? I will answer this question with four propositions that should guide an approach to the Fathers on this controversial issue of creation and evolution.

A RESPONSIBLE APPROPRIATION WILL
UNDERSTAND THE IMPORTANCE OF THE
CHURCH FATHERS FOR CHRISTIANS TODAY

This may seem peripheral to some readers, but it is not. While lip service is often paid to the Fathers' importance in Christianity, it is not always clear why some make these appeals. Judging by how they are appropriated in this controversy, it appears that the Fathers are used merely to add to the list of those who apparently held creation science conclusions based on Genesis 1. But if we do not understand the debt we owe to the early leaders of the Christian church, we will not see the need to understand how their perspective on this contentious issue may inform and enliven evangelicalism to a Christianity that is true to its heritage. In the first chapter, I outline how the Fathers played a significant and foundational role in orthodox Christian thought, including collecting the very documents we appeal to as sacred Scripture.

IT WILL HONOR PATIENT STUDY OVER
SIMPLISTIC CONCLUSIONS

Throughout the book, particularly chapter two, I outline some of the problems that occur when a less than deliberate and patient approach to the Fathers is employed. Patient study requires thoughtfulness and deliberation. Unfortunately, the way we evangelicals tend to deal with controversy is little different than the culture where these virtues are not cultivated. We prefer quick answers and easy conclusions rather than patient and humble deliberation. Reading the Fathers is not easy. It requires us to think differently. I will never forget the words of my PhD supervisor as we discussed a chapter submission of my dissertation. In the context of his frustration with some of my conclusions, he said, "Stop thinking like a Protestant!" My supervisor was a Protestant himself, so this was not a call to become a Roman Catholic. It was his way of telling me that I had to avoid forcing others into my own way of thinking or assuming that everyone thinks like me. To consider perspectives foreign to us, we may need to suspend conclusions and enter into a foreign way of thinking, which can be difficult and very uncomfortable. We need to let the church fathers speak for themselves rather than force our questions and controversies on them, as if they exist only to answer our modern concerns. A patient appreciation for these things will help us to hear them – not because they did not make mistakes but because they did not make the same mistakes we do. This will help direct us to the issues that matter in the controversy rather than those that may just be missing the point.

OUR EXPECTATIONS OF THE BIBLE WILL AFFECT HOW WE INTERPRET IT

This flows naturally from the previous point. Evangelicalism's historical development has shaped our concerns. In a context where the Bible was attacked, the Christian faith was increasingly marginalized, which led to a strategy of "proving" the historical veracity of the Bible over against its detractors. Context determined for us expectations from the Bible that were not necessarily paramount to previous Christian generations. This is not to say that previous generations did not affirm the veracity of the Bible. Rather, there were different expectations of how the Bible ought to function in the life of the church and its importance in pointing to the center of history – Jesus Christ.

The center of history for the early church was Jesus Christ. This meant that everything either pointed to him or grew from him. For early Christians, he was the subject of the Scriptures. They looked for him first and foremost. As we will see, the Fathers read this in the very first words of Bible, "In the beginning," and identified the beginning as Christ himself. They saw a distinct parallel between the opening of Genesis and the opening of John's Gospel. The parallel was not a mere historical connection but rather a spiritual connection implanted by the author of both texts – the Holy Spirit. Expectations of a merely historical connection or proof from the Bible flattens its importance in early Christianity and domesticates our expectations and how we interpret it.

The Fathers interpreted the Bible differently than we do because they had different expectations. At the very least, we would do well to patiently consider why this was so rather than allowing our modern historical sensibilities to rule spiritual and allegorical readings out of bounds before they are ever in play. We will see that dividing early Christian interpreters into either literalists or allegorists fails to accurately understand the context, and results in inaccurate, misleading, and naive conclusions about their readings of Genesis 1. To assume that the church fathers were simply precursors to our "proper" methods of grammatical-historical biblical interpretation robs them (and us!) of their potential to show us where we may need correction, advice, and guidance.

A RESPONSIBLE APPROPRIATION WILL PAY ATTENTION TO THE CONTEXT OF PATRISTIC WRITINGS

In what follows, I will comment many times on the dangers of proof-texting the Fathers. Proof-texting grows out of a failure to adequately consider context. This temptation appears when the desire to prove a point takes precedence over a responsible reading. It masquerades as such because it appeals to primary sources, but it does so in a way that is impatient and irresponsible.

For example, chapter four discusses how some have used Basil of Caesarea to champion a modern method of reading Scripture. But the broader context within which he ministered and the narrower context of his own writings show that using Basil in this way is much too simplistic and requires a deliberate proof-texting to prove a point rather than patiently deliberating on the meaning of his comments about allegory. This requires a contextual reading, which is what chapters five through seven demonstrate. This may take more time, energy, and patience, but if we are to bring our heritage to bear on an issue of importance to the present Christian community, we must make that commitment to responsible interaction with these sources. Those who have been called to this kind of ministry must model it carefully and with integrity.

A WORD ABOUT SOURCES

Biblical quotations are all from the New Revised Standard Version. My use of English translations of the Fathers requires comment. For their works specifically on Genesis, I have chosen to use the translations judged by the community of scholars as superior. English translations other than those on Genesis are also from ANF, NPNF[1], and NPNF[2].[7] If a translation appears in this collection, it is noted.

---

[7]See the Christian Classics Ethereal Library, www.ccel.org.

PART I

UNDERSTANDING

the

CONTEXT

1

WHO ARE *the* CHURCH FATHERS, *and* WHY SHOULD I CARE?

S everal years ago I was invited to give a presentation to an adult Sunday school class at a Mennonite church in my community. I called the presentation "Back to the Sources: An Introduction to the Great Thinkers of the Early Church" and was excited to share my passion for the church fathers with this audience. Unfortunately, my hearers did not share my excitement. At best they could not understand why we would need anything other than what we have in our Bibles. At worst, they could not understand why a good conservative Christian would recommend these figures from a church and an age that was, in their opinion, far from the purity of New Testament Christianity.

Granted, my experience above may be unique, but I doubt it. An argument could be made that the necessity of an introductory chapter in this book about the importance of the church fathers is a symptom of a greater problem within our churches that my experience illustrates. For reasons beyond the scope of this book, our own Christian heritage, which includes the church fathers, has been deemed, at best, marginally helpful for the twenty-first-century Christian. At worst, the history between the apostles and the Reformers has been judged as an era best left in the past because of its perceived distance from "true" Christianity. For many Christians the idea that we should appeal to the church fathers, who belong to that era, as part of our own Christian heritage is foreign, suspect, or even impious. The Christianity of that age has been seen as transitory, naive, and even problematic, and therefore an unnecessary resource for Christian faithfulness today. After all, wasn't it this kind of Christianity that the Protestant Reformers opposed?[1]

One important reason why some Protestants fail to recognize the connection we have with the unfolding story of the people of God in the early church is because of what D. H. Williams calls a "fall paradigm."[2] In this understanding, at some point after the apostolic age the church "fell" from its pure existence, and from this fallen condition the Roman Catholic Church emerged. This understanding renders the leaders, creeds and councils, and holy days of the ancient church suspect because they were either complicit in or products of this fall. It was not until the Reformation in the sixteenth century that true (New Testament) Christianity was rediscovered and "restored and set on its originally intended course."[3]

To many Protestants this is not merely a paradigm, but reality.[4] Yet for most it is implicitly assumed rather than explicitly affirmed. In *Retrieving the Tradition and Renewing Evangelicalism*, Williams argues persuasively that this way of interpreting history has its own history determined by certain motivations that gave rise to it in the first place. There

---

[1]D. H. Williams, "*Similis et Dissimilis*: Gauging Our Expectations of the Early Fathers," in *Ancient Faith for the Church's Future*, ed. Mark Husbands and Jeffrey P. Greenman (Downers Grove, IL: IVP Academic, 2008), 71.

[2]D. H. Williams, *Retrieving the Tradition and Renewing Evangelicalism: A Primer for Suspicious Protestants* (Grand Rapids: Eerdmans, 1999), esp. 101-32; and D. H. Williams, "Constantine, Nicaea and the 'Fall' of the Church," in *Christian Origins: Theology, Rhetoric and Community*, ed. Lewis Ayres and Gareth Jones (London: Routledge, 1998), 117-35.

[3]Williams, *Retrieving the Tradition*, 102.

[4]Paradigm = a constructed pattern or model.

may be better ways of looking at Christianity's history that can have corrective value for the fall paradigm. The fall paradigm automatically creates a barrier between the church today and our heritage because it assumes that we developed independently of, and even in spite of, the early post-apostolic church.

There are many reasons why the fall paradigm fails to convince some people of its viability.[5] First, it is an overly simplistic way of reading history. History tends to be messy, and a reading that simply categorizes into good and bad runs the risk of oversimplification and overgeneralization. Second, it does not accurately represent what many of the sixteenth-century Reformers believed about the ancient church and the Fathers. Many from within Protestantism itself are seeking to remind its adherents that the Reformers actually relied extensively on the sources of early Christianity because they saw themselves in continuity with the church fathers and the historic teaching of the church.[6] Reformation scholar David Steinmetz argues that the Reformers were very concerned that their teachings should match those of the Fathers. Steinmetz notes that they "turned to the Fathers because they found them important sources of insight into the text of Scripture."[7] The third, and perhaps most significant, problem with the fall paradigm is that it robs present-day Christians of their own heritage.

Calls for a recovery of the church fathers for the life of the church *today* have been underway within evangelicalism for many years now. Doctrine that we call orthodox, as well as the Bible to which we appeal for this essential doctrine, has deep roots in the mediating work of Christians after the apostles. Much of our understanding of what the Bible teaches has come to us through the church fathers. I would like to

---

[5]The reasons that follow are based on Bryan Litfin, *Getting to Know the Church Fathers: An Evangelical Introduction* (Grand Rapids: Brazos Press, 2007), 26-28. For a much more sustained and detailed explanation and critique, see Williams, *Retrieving the Tradition*; and Williams "Constantine, Nicaea and the 'Fall.'"

[6]This is essentially what I tried to do in Craig D. Allert, "What Are We Trying to Conserve? Evangelicalism and *Sola Scriptura*," *Evangelical Quarterly* 76 (2004): 327-48.

[7]David Steinmetz, "Why the Reformers Read the Fathers," *Christian History* 80, no. 4 (2003): 10.

unpack this as I make an apologetic for why we should at least consider the church fathers' interpretation of the six days of creation in Genesis.

WHO WERE THE CHURCH FATHERS?

The Greek and Latin terms for *father* are similar: in Greek it is *patēr* and in Latin it is *pater*. This is why the discipline that studies the church fathers is called *patrology*, or *patristics*, which means something like the teaching or study (*logos*) of the Fathers. One will often hear and read references to the patristic age. This simply means the age of the church fathers. But who were they?

The term *father* as a teacher, leader of a school of philosophers, or rabbi occurred in Jewish, Cynic, and Pythagorean circles.[8] In Christianity, *father* as an honorary title represents the "confluence of a host of common, human, OT, and Greco-Roman conceptions."[9] The understanding of a father as one who has gone before us is common in day-to-day experience. Although not everyone can claim to have had an ideal father, we can surely grasp the concept of one. He is someone who guides his children with wisdom gained from life experiences.[10] From here it is not a far step to the concept of a father as one who is a spiritual guide because of his experience in the faith and his responsibility and ability to hand on that faith.

In the New Testament the apostles saw themselves as fathers to the nascent church. In 1 Corinthians 4:14, Paul speaks of himself as a father to the Corinthian church.[11] He addresses both Timothy and Titus as loyal children in the faith.[12] John greeted his readers as "my children" and "my

---

[8]Christopher A. Hall, *Learning Theology with the Fathers* (Downers Grove, IL: InterVarsity Press, 2002), 19.

[9]Hubertus Drobner, *The Fathers of the Church: A Comprehensive Introduction*, trans. Siegfried S. Schatzmann (Peabody, MA: Hendrickson, 2007), 3.

[10]Litfin, *Getting to Know the Church Fathers*, 16.

[11]1 Cor 4:15: "For though you might have ten thousand guardians in Christ, you do not have many fathers. Indeed, in Christ Jesus I became your father through the gospel."

[12]2 Tim 1:2: "To Timothy, my beloved child"; Titus 1:4: "To Titus, my loyal child in the faith we share."

little children."[13] Peter even appealed to Christians of his own generation as "the fathers."[14] This kind of use extended into the patristic age. Writing in approximately 180 CE, Irenaeus claimed that "when any person has been taught from the mouth of another, he is termed the son of him who instructs him, and the latter [is called] his father."[15] Clement of Alexandria (ca. 150–ca. 215) says that "words are the progeny of the soul. Hence we call those who have instructed us, fathers. . . . And every one who is instructed, is in respect of subjection the son of his instructor."[16]

The term *father* in Christianity can therefore refer to previous generations of believers who have continued to guide their spiritual descendants in the church through history up to our own age.[17] This has prompted Christopher Hall to describe a father in the faith as "someone who is familiar with the teachings concerning the life and ministry of Jesus Christ and can be trusted to hand on faithfully and correctly the tradition that he himself has already received. Trustworthiness of character and rootedness in the gospel are non-negotiables in the life of a father."[18]

In this description it appears that anyone with similar qualifications could be labeled a church father. This is particularly true in the Greek Orthodox Church, of which Panagiotes Chrestou states:

> The Church has never excluded the appearance of renowned teachers in her bosom, who are outstanding bearers of the divine grace of the divine spirit, and she has never restricted this appearance to any particular period of her history. Orthodox ecclesiastical consciousness, which attributes the title of father in every epoch to elect vessels of grace that lived in previous epochs, has already pushed the patristic period to the end of the Byzantine era and is pushing forward beyond it more and more.[19]

---

[13]3 Jn 4; 1 Jn 2:1.

[14]2 Pet 3:4.

[15]Irenaeus, *Against Heresies* 4.41.2 (ANF 1:524).

[16]Clement of Alexandria, *The Stromata* 1.1.2 (ANF 2:299).

[17]Litfin, *Getting to Know the Church Fathers*, 17.

[18]Hall, *Learning Theology*, 19.

[19]Panagiotes K. Chrestou, *Greek Orthodox Patrology: An Introduction to the Study of the Church Fathers*, trans. George Dion. Dragas, Orthodox Theological Library 2 (Rollinsford, NH: Orthodox Research Institute, 2005), 15-16.

Even though Chrestou rightly recognizes, as we all should, the contribution of great leaders in the church throughout the ages, he "does not exclude the habit of regarding the Fathers of the first Christian centuries, when the foundations of the Christian institutions were first laid and the dogmatic teaching was specified to a large extent, as occupying a privileged position."[20] In agreement with Chrestou, the church fathers are limited to a particular age when foundational and baseline actions and beliefs were hammered out. Evangelical Christopher Hall recognizes the importance of Christian leaders in this age as those who preserve and hand on the teachings concerning Christ and his ministry (apostolic teaching), foundationally exemplified in "the conciliar decisions of key councils such as Nicaea (A.D. 325), Constantinople (381) and Chalcedon."[21]

There has not been complete agreement on limiting the age or even the orthodoxy of the church fathers. In his famous four-volume work on the Fathers, Johannes Quasten states that these early leaders include both orthodox and heretical writers up to Isidore of Seville (d. 636) in the West and John Damascene (d. 749) in the East.[22] Most, however, would not want to include as church fathers men who were deemed heretical. The dates assigned by Quasten are also somewhat arbitrary since most historians consider the seventh and eighth centuries as part of the early medieval age. For various reasons, I see a shift occurring with the death of Augustine (430) and the Council of Chalcedon (451). Therefore, I put the end of the patristic age around 451.

In 374 or 375, Basil of Caesarea included a list of church fathers that supported his argument in *On the Holy Spirit*.[23] The appeal to previous leaders in the church was important because it represented the

---

[20]Ibid., 16.

[21]Hall, *Learning Theology*, 19.

[22]Johannes Quasten, *Patrology*, vol. 1, *The Beginnings of Patristic Literature: From the Apostles' Creed to Irenaeus* (Westminster, MD: Christian Classics, 1986), 1.

[23]Basil, *On the Holy Spirit* 29 (NPNF² 8:45-48).

responsible safeguarding and handing on of the faith. Augustine did the same thing in his work against Pelagius. In about 434, a few years after the death of Augustine, a monk named Vincent of Lérins (d. before 450) wrote a document called *Commonitory* (meaning something like "an aid to memory"), a classic formulation of what has contributed to the main criteria by which we identify a church father.[24] Building on Scriptures that appeal to the wisdom and guidance of past leaders in the church,[25] he states that his purpose in this aid to memory is to "put down in writing the things which I have truthfully received from the holy Fathers."[26] It is Vincent's conviction that the Fathers, along with Holy Scripture, are invaluable for Christians to distinguish truth from heresy. Thus, he counsels that

> we hold that faith which has been believed everywhere, always, by all. . . . This rule we shall observe if we follow universality, antiquity, consent. We shall follow universality if we confess that one faith to be true, which the whole Church throughout the world confesses; antiquity, if we in no wise depart from those interpretations which it is manifest were notoriously held by our holy ancestors and fathers; consent, in like manner, if in antiquity itself we adhere to the consentient definitions and determinations of all, or at the least of almost all priests and doctors.[27]

Vincent's admiration of these "holy Fathers" has influenced a set of four criteria that has emerged, wherein a church father is thought of as one who is ancient, orthodox in doctrine, holy in life, and approved by the church.[28]

---

[24]Vincent of Lérins, *A Commonitory: For the Antiquity and Universality of the Catholic Faith Against the Profane Novelties of all Heresies* (NPNF[2] 11:123-59).

[25]As cited by Vincent in *Commonitory* 1.1 (NPNF[2] 11:131). See also Deut 32:7, "Ask thy fathers and they will tell thee, thine elders and they will declare unto thee"; Prov 22:17, "Bow down thine ear to the words of the wise"; and Prov 3:1, "My son, forget not these instructions, but let thy heart keep my words."

[26]Vincent, *Commonitory* 1.1 (NPNF[2] 11:131).

[27]Vincent, *Commonitory* 2.6 (NPNF[2] 11:132).

[28]The criteria are well-known and used, especially in the West, and can be found in any introduction to the church fathers. The brief descriptions employed here draw on Drobner, *Fathers of the Church*, 4.

Ancient. A Father must belong to the period of the ancient church. The difficulty discussed above concerning the closing date of the age of the Fathers need not concern us here.

Orthodox in doctrine. A church father had to be in agreement with the common teaching of the church as represented in Scripture, the rule of faith, and the creeds. Contrary to some popular belief, appeal to the Fathers does not rest in an understanding of them as inerrant: not everything they said or wrote is understood as binding. Yet there is an aspect of their body of writing that has been recognized as resting within the bounds of correct teaching.

Holy in life. Here it is important to realize that the Fathers were "intensely human"[29] and therefore susceptible to the same temptations, flaws, and shortcomings as we are in our own quest for a holy life. This means that in the zeal for their deep-seated commitment to the things of God they may exhibit some aspects that are "less laudable."[30] But for all their faults, they were committed to the gospel: "They lived and breathed the Scriptures. And many willingly laid down their lives for the sake of Christ. . . . We would err if we allowed their weaknesses to blur or block the significant contributions their thoughts and lives can give the church today."[31]

Approved by the church. This means that the church at large recognizes the importance of these figures. This recognition was not necessarily explicit, as in a council. Rather, it is recognition of the person and writings through reference and affirmation.

WHY SHOULD I CARE?

This is likely a much more significant issue to my intended audience than the question of who the Fathers were. Therefore, here it would be appropriate to outline the numerous ways in which the Fathers are important in Christianity, past and present.

---

[29]Christopher A. Hall, *Reading Scripture with the Fathers* (Downers Grove, IL: InterVarsity Press, 1998), 51.

[30]Boniface Ramsay, *Beginning to Read the Fathers* (Mahwah, NJ: Paulist Press, 1985), 5.

[31]Hall, *Reading Scripture*, 52.

The church fathers help us remember who we are. In the 2002 movie *The Bourne Identity* Matt Damon's character Jason Bourne wakes up floating in the ocean with two gunshot wounds in his back.[32] He is rescued but cannot remember anything about his past, and he soon realizes that he is fluent in several languages and capable of advanced combat skills. Bourne's amnesia means that he has no memory of how he ended up floating in the ocean, how he received the gunshot wounds, or how he attained his advanced skills. This provides the motivation for Bourne to seek out his identity in order to understand who he is and how he attained these abilities.

Baptist theologian and patristics scholar D. H. Williams argues that many evangelicals are like Jason Bourne.[33] The amnesiac has a history, but it cannot be recalled. All is present and future; there is no past. Important events, rites, experiences, and relationships are simply not available for the sufferer to relate to the present and the future. For Williams, the result in some evangelical circles has been a perspective on history and theology that views these as irrelevant for today's Christian. Pragmatism has largely overtaken any need for thoughtful and patient reflection about how our past might affect our future, because we simply do not know our past. "It is not," states Williams "that Christians are purposely ignoring Paul's final words to Timothy, 'preserve the pattern of sound teaching . . . guard the good deposit that was entrusted to you,' it is that they are no longer sure what this 'deposit' consists of, or where it can be found. In some cases, finding this 'deposit' does not matter anymore."[34]

Again, it is *not* that evangelicals do not have a history. Just as Jason Bourne had language and combat skills he could not explain, evangelicals

---

[32]Robert Ludlum, Tony Gilroy, and William Blake Herron, *The Bourne Identity*, directed by Doug Liman (Universal City, CA: Universal Pictures, 2002).

[33]Williams, *Retrieving the Tradition*, 9-39. Williams does not use *The Bourne Identity* as an example. My point here is that the amnesia suffered by Jason Bourne is illustrative of the amnesia indicated by Williams.

[34]Ibid., 10.

hold certain essential doctrines and perform certain important rites, passed down to us from the Fathers, that they cannot explain historically and, by implication, theologically. There are remnants of the church fathers in our evangelical churches, but we fail to see them or understand their influence.

> Telltale signs of the Patristic Tradition can still be found within evangelical churches: baptism in the name of the Trinity, Christ admitted as fully human yet worshipped as God, occasional acknowledgement of the Apostles' or Nicene Creed, and, more fundamentally, the authoritative use of a collection of documents known as the New Testament. But these vestiges of the early faith are just that, *vestige*, i.e., footprints or tracks that speak of a doctrinal and confessional past which has been peripheral for so many evangelicals that it has ceased to guide the direction of many present-day congregations and in some cases, is forgotten. There is a shared sense that the central elements of the Christian faith must be preserved in the Church, but it is not clear why or what practical purpose they serve for the present needs of everyday ministry.[35]

Later, I will direct my attention to the connection of these essential elements of the Christian faith to the church fathers. Right now my focus is on recognizing that who we are as Christians requires a knowledge of where we come from – we need to remember.

Williams's point about amnesia in evangelicalism is shown, quite ironically, in a recent and fairly positive review of his book on the internet blog of student ministries coordinator and blogger Meriah, who identifies herself as a "North American Baptist." This locates her in the same denominational orbit as Williams and within the evangelical tradition. At several places in her review Meriah acknowledges her ignorance of history and tradition in her church and in her own thinking. In one particularly telling section, Meriah confesses the following:

> As I read through Williams' book I caught myself on the defensive side and constantly arguing with any point of contention he had with the Evangelical Church. I hated that he was commenting negatively on the

---

[35]Ibid., 11.

way I had always practiced communion. I felt like he was personally attacking everything I have ever known about my faith. It was not until I became aware of this bias that I actually started to process what Williams was saying.

In all honesty, *I do not know why Baptists believe in a symbolic form of communion instead of transubstantiation or consubstantiation. I do not know why Baptists choose to baptize believers and "dedicate" infants. I do not know why our church has a mission statement that the congregation is not even familiar with and why we do not recite or know any creeds like other denominations.*[36] (italics added)

Meriah confesses that this ignorance bothers her and that she actually wants to know more. But even in her self-confessed ignorance — not only of her own denomination's history but also of broader Christian history — she asserts that there are certain early church practices she would not like to see "reinstated." Her reason is shockingly simplistic and troubling: "I am part of a Baptist congregation for a reason. For example, although I do not know the Tradition behind it, I love that Baptists baptize believers." There is something deeply troubling about an active minister of the gospel whose "reason" for adhering to baptism of believers is merely because she "loves it." Even after reading a book about how the evangelical tradition is anchored in the past, this reviewer cannot understand the point Williams is making, much less appropriate that past for present ministry. It appears this amnesiac has no desire to discover who she really is as a Baptist or as a believer in Christ.

This sort of "willful amnesia," according to Robert Wilken, is "a self-imposed affliction that would rob us of our lives of depth and direction."[37] It robs Christians of their own rich heritage of theology, worship, and devotion to Christ. Without any obligation to the past, autonomy may rule under the guise of "biblical Christianity." The connection of appropriating

---

[36]Meriah, "Reflection on D. H. Williams' Retrieving the Tradition & Renewing Evangelicalism: A Primer for Suspicious Protestants," October 19, 2014, http://meriahtigner.blogspot.ca/2014/10/reflection-on-dh-williams-retrieving.html?view=magazine.

[37]Robert L. Wilken, *Remembering the Christian Past* (Grand Rapids: Eerdmans, 1995), 170.

the Christian past with biblical Christianity may sound strange to some, but the confusion can be cleared up in the next reason for why we should care about the church fathers.

Our New Testament is a legacy from the church fathers. It is impossible to understand the development and function of the New Testament in the early church without understanding the context in which it was set. When the church was stripped of its Jewish framework in the wake of Jerusalem's destruction in 70 CE, it was forced to identify itself apart from the Jewish framework in which she was birthed.[38] In this process of self-identification, there were several things that the church developed to help it find stability and growth in the first few tumultuous centuries of her existence. The development of church leadership, increased attention to universal creeds, and the formation of a specifically Christian canon of Scripture all evolved interdependently, and a contextual understanding of them shows us the important role the church fathers played in the formation of our New Testament.

There is an understandable desire for some in the evangelical tradition to locate the closing of the New Testament as early as possible in the church's history. This is because of our Protestant distinctive, *sola Scriptura* or "Scripture alone." If, as the reasoning goes, the Bible is as central to the Christian faith as Protestants believe (and it is!), and it is all we need (is it?), then it must have been a very early component of historic Christian orthodoxy. This is the guiding assumption with which some approach the history of assembling the New Testament. But there is a chronological problem with the guiding assumption – the closed New Testament canon was not a particularly early feature of historic Christian orthodoxy.[39]

---

[38]A very helpful summary of this is found in Mark Noll, *Turning Points: Decisive Moments in the History of Christianity*, 2nd ed. (Grand Rapids: Baker Academic, 2000), 23-46.

[39]A more detailed examination of this is found in Craig D. Allert, *A High View of Scripture? The Authority of the Bible and the Formation of the New Testament Canon*, Evangelical *Ressourcement* Series (Grand Rapids: Baker Academic, 2007). See also Harry Y. Gamble, *The New Testament: Its Making and Meaning* (Eugene, OR: Wipf & Stock, 2002); and Lee M. McDonald, *The Biblical Canon: Its Origin, Transmission, and Authority* (Peabody, MA: Hendrickson, 2007).

Generally speaking, a canon is anything that functions as a rule or norm – a standard against which something is measured. Today, when Christians speak of "canon" it is usually in reference to that closed collection of authoritative texts. But its initial use in the church had nothing to do with texts. The apostle Paul used the term *kanōn* four times, and none of them is a reference to a collection of writings.[40] Once he uses it to refer to the standard of Christian behavior,[41] and the three other times (all in the same passage) in connection with the sphere that God has given to Paul for his missionary work.[42]

For Irenaeus of Lyons (130–202), *canon* meant the rule of faith – the content of essential Christian belief. This was also true of other early church fathers.[43] The word soon moved from this more fluid usage to refer to concrete things like decisions of councils,[44] monastic rules,[45] clergy,[46] and finally to a list, index, or table – something to which a person can orient oneself. This shows the natural growth of the category called canon to include an official list of Christian Scripture *alongside* these other canons in the church.

Historians of the New Testament canon often consider "criteria of canonicity," which constitute a retrospective scheme to help us understand

---

[40]The following comments on definition and use of canon in early Christianity are from Bruce M. Metzger, *The Canon of the New Testament: Its Origin, Development, and Significance* (Oxford: Clarendon Press, 1987), 289-93.

[41]Gal 6:16: "As for those who will follow this rule [*kanōn*] – peace be upon them, and mercy, and upon the Israel of God."

[42]2 Cor 10:13-16: "We, however, will not boast beyond limits, but will keep within the field [*kanōnos*] that God has assigned to us, to reach out even as far as you. For we were not overstepping our limits when we reached you; we were the first to come all the way to you with the good news of Christ. We do not boast beyond limits, that is, in the labors of others; but our hope is that, as your faith increases, our sphere [*kanōna*] of action among you may be greatly enlarged, so that we may proclaim the good news in lands beyond you, without boasting of work already done in someone else's sphere [*kanōni*] of action."

[43]See, e.g., Clement of Rome, *1 Clement* 7.2; Clement of Alexandria, *The Stromata* 4.15.98, 6.15.124; Hegesippus, cited in Eusebius, *Church History* 3.23.7; and Polycrates of Ephesus, cited in Eusebius, *Church History* 5.29.6.

[44]The decrees from councils came to be known as "canons."

[45]On the use of "canon" to refer to monastic rules, see Athanasius, *On Virginity* 12; Basil of Caesarea, *Detailed Rules* 45.1; and Gregory of Nazianzus, *Epistle* 6.

[46]The canons of the Council of Nicaea (325), for example, used the word to refer to an official list of clergy who were attached to a particular church. The clergy who were on this list were referred to as "those in the canon." See *Canons* 16, 17, and 19 in NPNF[2] 14:35, 36, 40.

why some orthodox Christian documents came to be valued above other equally orthodox Christian documents. The scheme is devised through a contextual reading of the leaders of the early church and their use of these numerous documents. Although order and ranking in importance differs depending on the historian, there are four such criteria: apostolicity, orthodoxy, catholicity, and widespread use. For the purpose of this section, I will single out the criterion of orthodoxy – the congruity of a document's content with the faith or teaching of the apostles.

If orthodoxy was a criterion for potential inclusion of documents in a closed New Testament, then there had to be an already existing standard of orthodoxy against which the documents could be measured. In many of the Christian writings that were eventually included in the New Testament, we can actually see this appeal. There is an explicit appeal to and exhortation for the believer to remain in and hold to the faith that the church had received. There is a progression of thought that moves from the teaching of Jesus, who hands it over to his apostles, who subsequently pass it on to the leaders of the church who are charged with guarding the pure teaching.

So when Paul urges the Galatian believers to avoid any gospel "contrary to what you received" (Gal 1:9), he is appealing to this process as authoritative and proper. A similar appeal is found in the exhortation to the Thessalonian church when the believers are urged to "stand firm and hold fast to the traditions" (2 Thess 2:15) taught to them and to stay away from people who do not live "according to the tradition . . . received from us" (2 Thess 3:6). What we see in these and other passages is an appeal to orthodoxy.[47]

Since these appeals appear in the documents that were not yet part of a New Testament, they cannot have been made to the New Testament. There existed an accepted and known standard to which these passages were referring. Similar appeals to orthodoxy occurred into the second

---

[47]Col 2:6, 8; 1 Tim 4:6; 2 Tim 1:13-14; 3:8; 4:7; Titus 1:9; 2:1; 1 Jn 1:5; Jude 3, 20.

century, using phrases like "the word of truth," "apostolic teaching," "tradition," "sound doctrine," and "the faith." These appeals to orthodoxy can be understood as appeals to something called the rule of faith.

Scholars define the rule of faith in different ways, but even in these differences there is a remarkable similarity in content and function. One scholar calls it a "graph of the interpretation of the Bible by the Church of the second and third centuries."[48] Another defines it as "a sort of communal linguistic awareness of the faith delivered to the apostles, which sufficed the church for generations."[49] The rule of faith guided missionary proclamation, shaped teaching, identified heresy, and functioned in the life of the church whenever a brief statement of the gospel's content was necessary.[50] Generally speaking, the rule of faith is a fluid expression of orthodox Christian belief, including affirmation of one God the Creator, Jesus Christ and his coming, the Holy Spirit, and future judgment – all what we would call essentials of the faith.

Irenaeus provides a good example of how the rule of faith functioned in a symbiotic relationship with church leadership and the documents that were eventually included in the New Testament canon. Gnostic Christians in Irenaeus's day were arguing that they had secret revelations passed on to them from the apostles themselves. Irenaeus's response was that this is impossible because the "certain gift of truth," taught by the apostles, is maintained and passed down through the historic leadership in the church.[51] It is, he argues, known everywhere and protected in its purity by the church leaders: "For this gift of God has been entrusted to the Church, as breath was to the first created man, for this purpose, that all the members receiving it may be vivified."[52] This gift of truth is, of course, the rule of faith, of which Irenaeus gives us a typical example:

---

[48]R. P. C. Hanson, *Tradition in the Early Church* (London: SCM Press, 1962), 127.
[49]Robert W. Jenson, *Canon and Creed*, Interpretation (Louisville, KY: Westminster John Knox Press, 2010), 15.
[50]Ibid.
[51]Irenaeus, *Against Heresies* 4.26.2 (ANF 1:497).
[52]Ibid., 3.24.1 (ANF 1:458).

The Church, though scattered through the whole world to the ends of the earth, has received from the Apostles and their disciples the faith in one God, the Father almighty, who made the heaven and the earth, and the seas, and all that in them is; and in one Christ Jesus, the Son of God, who became flesh for our salvation; and in the Holy Ghost, who through the prophets preached the dispensations and the advent, and the birth from the Virgin, and the passion, and the resurrection from the dead, and the bodily assumption into heaven of the beloved Christ Jesus, our Lord, and his appearing from heaven in the glory of the Father, to comprehend all things under one head, and to raise up all flesh of all mankind.[53]

The rule of faith thus represents a certain consistency of belief, a universal acceptance that guarantees the maintenance of the true faith in contrast to the secrecy of the Gnostics.

Irenaeus shows us that the second-century church had a reliable standard (canon) of orthodoxy that had been faithfully handed down (the literal meaning of "tradition") from the apostles and safeguarded by the leaders of the church. The term *canon* did not come to be used as an appellation for a collection or list of Christian writings (NT) until the mid-fourth century. Earlier in the second century the word designated what the church acknowledged as having regulative authority for its faith and life, the rule of faith. It was only after Athanasius (296–373) that the term regularly came to denote a closed collection of authoritative writings to which nothing could be added or from which nothing could be taken away.[54] During the second century the church answered the Gnostic challenges with the rule of faith, which was then defended by an appeal to the apostolic writings, many of which would eventually become part of the New Testament.[55]

The same kind of thing occurred into the third century, as is shown by Serapion of Antioch (d. 211). Eusebius's *Church History* tells us that Bishop Serapion wrote a book to refute the teachings of a document

---

[53]Ibid., 1.10.1 (ANF 1.330-31). See also other explicit delineations of the content of the rule in *Against Heresies* 1.9.4; 3.4.1-2; 5.20.1.

[54]See Athanasius, *Letter 39* in NPNF² 4:551-52.

[55]See Irenaeus, *Against Heresies* 1.8.1; 1.9.1-4; and Tertullian, *The Prescription Against Heretics* 8-9.

called the Gospel of Peter.[56] A church under his jurisdiction in Rhossus, Syria, had received the Gospel of Peter as if it was written by the apostle Peter and was using it in their worship and teaching. Serapion had not read the book but assumed that if the church at Rhossus was using the document, they must have discerned it properly since they "held the true faith." When Serapion actually read the document, he was led to write a book against it because of its docetic doctrine – it taught that Jesus did not have a real human nature, thus denying the "true faith."

What is interesting and informative here is that Serapion showed little concern for what was in the New Testament and what was not. In fact, neither the church at Rhossus nor Serapion showed an awareness of a closed New Testament canon. The silence of an appeal to it is deafening. According to Serapion, the teaching that Jesus was not truly human was expressly denied in the "true faith." Where is this true faith located? Yes, it is in the Scriptures that eventually made up our written New Testament canon, but it was also located in the rule of faith that served as the norm against which Serapion measured the Gospel of Peter before the New Testament canon existed – the rule of faith stating that Jesus "became flesh for our salvation."[57] The issue was dealt with on the basis of an appeal to orthodoxy that is represented in the rule of faith, not canonicity.[58] In this way, it was functioning as guardian of doctrine and the proper interpretation of Scripture.

This does not mean that Christian writings (both canonical texts and other orthodox documents) were unimportant or irrelevant during the second and third centuries. Quite the opposite, they were a vital part of a canonical heritage. But before there was a New Testament canon, there existed a rule of faith that, along with the leaders of the early church, functioned as the guardian of proper doctrine.

---

[56]Eusebius of Caesarea, *Church History* 6.12.1-6 (NPNF[2] 1:257-58).

[57]Irenaeus, *Against Heresies* 1.10.1 (ANF 1:330). See also *Against Heresies* 3.4.1-2; 4.33.7; and Tertullian, *On the Veiling of Virgins* 1; *Against Praxeas* 2; *Prescription Against Heretics* 13.

[58]L. M. McDonald, "The Integrity of the Biblical Canon in Light of Its Historical Development," *Bulletin for Biblical Research* 6 (1996): 118-19.

Irenaeus likens the church's guarding apostolic teaching to a rich man's depositing money in a bank.[59] Only the true church is in possession of these sacred truths in the canonical heritage. People who pervert these truths or add to them are thieves and robbers. At one point Irenaeus even claims that those who might not have access to the Christian Scriptures are still well served because the essentials of apostolic teaching are preserved and handed down by the leaders in the churches. This serves missionary efforts as well as attempts to ward off dangerous teaching.[60] It was this rule of faith, safeguarded by the church fathers, against which everything was measured, even the writings of the developing New Testament. We must be careful, however, not to press this too far.[61] In other words, we must not – in view of the importance of both the rule of faith and the church leadership by the Fathers – take this to mean that Christian writings were relatively unimportant in the early church. In fact, the later development of a collection of these writings speaks volumes to the contrary. But placed in proper context and chronology, we should not miss the important role played by the Fathers in this regard.

Even after the closing of the New Testament canon, the same kind of appeal was being made to the symbiotic relationship between rule of faith, church leaders, and Scripture.[62] In 434 Vincent of Lérins claimed that "the canon of scripture is complete."[63] He anticipates his readers' queries about the need for anything beyond Scripture:

> "Since the canon of Scripture is complete, and sufficient of itself for everything, and more than sufficient, what need is there to join with it the authority of the Church's interpretation?" For this reason, – because, owing

---

[59]Irenaeus, *Against Heresies* 3.4.1 (ANF 1:416).

[60]Ibid., 3.4.2 (ANF 1:417).

[61]Campenhausen very succinctly expresses the proper balance between Scripture and the rule of faith. See H. F. von Campenhausen, *The Formation of the Christian Bible*, trans. J. A. Baker (London: Adam & Charles Black, 1972), 329.

[62]For argument's sake I will assume a closing near the end of the fourth century even though there was still fluidity into the ninth century and beyond. See Allert, *High View of Scripture?*, 87-130.

[63]Vincent of Lérins, *Commonitory* 2.5 (NPNF[2] 11:132).

to the depth of Holy Scripture, all do not accept it in one and the same sense, but one understands its words in one way, another in another; so that it seems to be capable of as many interpretations as there are interpreters.[64]

The appeal here to the "Church's interpretation" is to the canonical heritage I have described above. It was needed before the closing of the canon, and it was needed after it – and that need continues today.

For Vincent, since the Scriptures are easily manipulated, a guide needs to be offered so that the faithful might be able to discern poor from proper interpretation. Heretics constantly appeal to Scripture. He states that there is "an infinite heap of instances, hardly a single page, which does not bristle with plausible quotations from the New Testament or the Old."[65] Why, Vincent asks, should we cast off the ancient faith of the church for the innovations of the heretics? The heretic would answer, "For it is written," and proceed to offer proofs from the apostles and the prophets.[66] In these cases Vincent gives an ancient and proven answer: Christians are to pursue that course which was commended to them by holy and learned men – the Fathers – and they must interpret the Bible in keeping with the canonical heritage.[67]

The canonical heritage contains the ways that the early church "laid down the baseline of essential Christian truths: confessions, creeds, doctrines, interpretations of the Bible, hymns, and so on."[68] The canon of Scripture was never meant to replace this canonical heritage but rather to act as an embodiment of the canonical tradition of the church. In this canonical heritage the church fathers were central.

Both Irenaeus and Vincent called on some aspect of our canonical heritage to safeguard apostolic teaching and proper interpretation. This did not happen in isolation, or alone, but within the very heart of the

---

[64]Ibid.

[65]Ibid., 25.64 (NPNF[2] 11:150).

[66]Ibid., 26.69 (NPNF[2] 11:151).

[67]Ibid., 27.70 (NPNF[2] 11:152).

[68]D. H. Williams, *Evangelicals and Tradition: The Formative Influence of the Early Church*, Evangelical *Ressourcement* Series (Grand Rapids: Baker Academic, 2005), 56.

church with the Fathers leading the charge. This is why the proper home of the Bible is the church and why Tertullian (ca. 155–240) refused to argue Scripture with the heretic. The heretic, he insists, had no right to claim Scripture, because it was not his own – by offering his unique, individual slant on interpretation, the heretic was interpreting it outside its true home.[69]

The significance for this is that by accepting the Bible as authoritative we must also accept the process and means through which it came to be – we are operating within the church's recognition of canon as operating within this broader context in which the Fathers, as leaders of the church, were major players.

ESSENTIALS OF CHRISTIANITY

The church fathers offer Christians today a "unique and privileged look at the ancient Church."[70] It is the church that emerged from the age of the apostles and began to develop what has come to be called the apostolic tradition. The historical proximity of the Fathers to the apostles leads Christopher Hall to speak of "hermeneutical proximity."[71] While this is certainly not a guarantee of infallible interpretations of the Bible, at the very least it allows us to read the Bible with the help of those who were much closer to the age of the apostles than we are. This can keep us mindful of our own culturally specific readings of Scripture and the danger of equating those with the Bible itself. Of course, the Fathers also had culturally specific readings, but that is just the point.

This is similar to the exhortation of C. S. Lewis over seventy years ago. In a wonderful little essay titled "On the Reading of Old Books," Lewis bemoans the neglect of ancient books because they are deemed irrelevant to our own times.[72] He claims that this neglect of ancient books

---

[69]Tertullian, *Prescription Against Heretics* 15.

[70]Ramsay, *Beginning to Read*, 17.

[71]Hall, *Learning Theology*, 21; and Hall, *Reading Scripture*, 38-41, 54.

[72]C. S. Lewis, "On the Reading of Old Books," in *God in the Dock: Essays on Theology and Ethics*, ed. Walter Hooper (Grand Rapids: Eerdmans, 1970), 200-207.

is nowhere more prevalent than in theology, and urges his readers to avoid the mistake of preferring new books to ancient ones. Of the several reasons he offers for the reading of old books, one is particularly relevant to our appropriation of the church fathers: "Every age has its own outlook. It is especially good at seeing certain truths and especially liable to make certain mistakes. We all, therefore, need the books that will correct the characteristic mistakes of our own period. And that means the old books."[73] For Lewis this is no utopian vision of an unreachable golden age. He knows that there is nothing magical about the past – "People were no cleverer then than they are now; they made as many mistakes as we. But not the *same* mistakes."[74] For Lewis, a very important way of avoiding a myopic understanding of our own theories and conclusions is to examine how our forefathers in the faith understood things. Recognizing that we are not the first to struggle to read Scripture in order to live faithfully before God can open up a whole new repository of guidance. It can allow us to see ourselves as situated within an ongoing tradition that extends from the past and speaks into our own lives.[75] Christopher Hall agrees: "If we rely solely on modern commentaries and systematic theologies, we might well overlook wisdom, patterns, concerns and models that can supplement and correct the insights offered by modern theological reflection."[76]

Lewis's advice is especially pertinent to those who affirm the authority of sacred Scripture, something that evangelicals hold in common with the church fathers. Understanding that we inhabit a long heritage manifested in persons who can help guide us in our Christian faith can be a tremendous help and blessing. Unfortunately, we sometimes don't make that important connection and we intentionally cut ourselves off from our own heritage. The reason is that our own

---

[73]Ibid., 202.
[74]Ibid.
[75]Stephen Fowl, introduction to *The Theological Interpretation of Scripture: Classic and Contemporary Readings* (New Malden, MA: Blackwell, 1997), xvii.
[76]Hall, *Learning Theology*, 21.

Christian history is viewed as somehow tainted or corrupt and, therefore, must be overcome in order to find "true Christianity." This is a pattern of thinking about history called restitutionism, which

> rejects traditional pre-modern history in order to restore "true history" and locates "true history" not in a tradition or the mystery of the church but in a lost yet supposedly recoverable body of "facts." It assumes that one group or person can be closer than another (corrupted) group or person to the original Jesus or the true Jesus or the true Paul solely by studying the documents of the New Testament.[77]

When Protestants couple a misunderstanding of the purpose of the Reformation with a doctrine of *sola Scriptura*, they may easily become susceptible to a restitutionist view of their own history.[78] Restitutionism construes the Protestant Reformation not as a reform of what went before but as a retrieval of a totally lost "true Christianity." This allows people to either bypass the ages in between the apostles and the Reformers or to very selectively choose events and figures according to restitutionist presuppositions. The result is a sharp dualism between Christianity in the patristic age and that of the "true" post-Reformation church.[79]

Restitutionists are suspicious of institutions, and this is what they see in the church of the patristic age. But this suspicion causes problems as any movement grows.[80] For the first generation of restitutionists, a movement can live in the enthusiasm of newness and the excitement of the Spirit. But the newness and excitement, not to mention mission carrying, is difficult to maintain in the second generation without institutions. It is very difficult to keep the first generation alive outside of institutions. Accepting the reality of the second generation means properly discerning the need for institutions while also recognizing

---

[77]Dennis D. Martin, "Nothing New Under the Sun? Mennonites and History," *Conrad Grebel Review* 5 (1987): 5.

[78]For a more detailed discussion of this, see Allert, "What Are We Trying to Conserve?"

[79]Williams, *Retrieving the Tradition*, 115.

[80]Martin, "Nothing New Under the Sun?," 8-9.

their vulnerability to corruption. So rather than a restitutionist outlook on history, some call for a reform approach to history that

> accepts the past all the way down to the present while at the same time calling for reform of institutions where they have become deformed. . . . It is an attitude that accepts institutions and has a basic attitude of trust toward the handing down (tradition) of institutions, even when it recognizes that institutions are deformed and in need of reform."[81]

The reform approach is illustrated in Alfred the Great's (849–899) preface to his translation of Gregory the Great's (ca. 540–604) famous *Pastoral Rule*:

> Our forefathers, who formerly held these places, loved wisdom, and through it they obtained wealth and left it to us. In this we can still see their tracks but we cannot follow them, and therefore we have lost both the wealth and the wisdom, because we would not bend with our minds to the track.[82]

The idea of "forefathers" means those who have gone *before* us. They have gone ahead of us even though they are *behind* us. Alfred uses the image of a track on which these forefathers have gone ahead. Christ is the trailblazer who is followed by the apostles and the church fathers. But, to our modern eyes, the trail leads "backward." Alfred explains that the task of Christian leaders is to keep that path open, to keep Christ's tracks visible. Christ, the apostles, and the Fathers, even though they lived centuries ago, are still out in front of us and leading us. We do not live in the age of the pioneer, because the trail has already been blazed.

The late Robert Webber argued passionately for the same kind of thing with his insistence that "the road to the future runs through the past."[83] To move "forward" requires looking "backward" toward Christ,

---

[81]Ibid., 11.

[82]Example used in Peter C. Erb, "Patristic and Medieval View," in *Perspectives on the Nurturing of Faith*, ed. L. Harder, Occasional Papers, no. 6 (Elkhart, IN: Institute of Mennonite Studies, 1983), 103.

[83]Robert E. Webber, *The Divine Embrace: Recovering the Passionate Spiritual Life* (Grand Rapids: Baker, 2006), 8. This is the overarching theme of Webber's Ancient-Future Series published by Baker, of which *The Divine Embrace* was the final installment. The others are *Ancient-Future Faith:*

the apostles, and the Fathers. This can be accomplished only by inhabiting the history that hands them over to us. When we read and study the Bible, worship in our churches, hear sermons, and participate in the rites of Communion and baptism (even as witnesses), we must recognize their foundation in our past and the institutions that have handed them on to us – the church and her leadership. This should drastically alter a rejection of the past merely because it is located in an institution. History is not understood as a sequence from beginning, through the middle, and to the end. Rather, it is a present memory, an actual recalling that retains its vital importance.[84]

The idea of reform rather than restitution thus changes the way we view our own past. The past is not just ages *passed*; rather, we inhabit it in and through our community of faith. It does not gloss over sin and corruption in our history – it believes that God has always been with his church, even when it was corrupted ("On this rock I will build my church, and the gates of Hades will not prevail against it," Mt 16:18). It does not seek to excise entire chunks of our history, because doing so would question God's presence in his church during these times of difficulty and even unfaithfulness. The "reform attitude accepts limitations of the history of an embodied church."[85]

It may appear justified to reject institutions that lapse into unfaithfulness and corruption, but it may be that the reaction itself should be reconsidered. Perhaps a better way to understand unfaithfulness in Christianity's history is to notice that the "occasion (although not the source) of temptations is first of all the good itself."[86] Perhaps we can

---

Rethinking Evangelicalism for a Postmodern World (1999); *Ancient-Future Evangelism: Making Your Church a Faith-Forming Community* (2003); and *Ancient-Future Time: Spirituality Through the Christian Year* (2004). See also *The Younger Evangelicals: Facing the Challenges of the New World* (Grand Rapids: Baker, 2002).

[84]Martin, "Nothing New Under the Sun?," 12.

[85]Ibid., 13.

[86]Gerald Schlabach, "Deuteronomic or Constantinian: What Is the Most Basic Problem for Christian Social Ethics?," in *The Wisdom of the Cross: Essays in Honor of John Howard Yoder*, ed. Stanley Hauerwas, Chris K. Huebner, Harry Huebner, and Mark Thiessen Nation (Grand Rapids: Eerdmans, 1999), 450.

learn something from an event in the life of ancient Israel in Deuteronomy 6–9, which relates a temptation that arose precisely *because* they were God's chosen people. Deuteronomy 8:7-17 states,

> For the LORD your God is bringing you into a good land, a land with flowing streams, with springs and underground waters welling up in valleys and hills, a land of wheat and barley, of vines and fig trees and pomegranates, a land of olive trees and honey, a land where you may eat bread without scarcity, where you will lack nothing, a land whose stones are iron and from whose hills you may mine copper. You shall eat your fill and bless the LORD your God for the good land that he has given you.
>
> Take care that you do not forget the LORD your God, by failing to keep his commandments, his ordinances, and his statutes, which I am commanding you today. When you have eaten your fill and have built fine houses and live in them, and when your herds and flocks have multiplied, and your silver and gold is multiplied, and all that you have is multiplied, then do not exalt yourself, forgetting the LORD your God, who brought you out of the land of Egypt, out of the house of slavery, who led you through the great and terrible wilderness, an arid wasteland with poisonous snakes and scorpions. He made water flow for you from flint rock, and fed you in the wilderness with manna that your ancestors did not know, to humble you and to test you, and in the end to do you good. Do not say to yourself, "My power and the might of my own hand have gotten me this wealth."

The Deuteronomist questioned neither God's desire to give the once oppressed people a land in which to securely prosper nor Israel's identity as the people whom God had called into covenant. Still, this moment of God's great gift came with a very great temptation and danger. On the day when they had fully entered the land and appropriated God's great gift to them, they were warned not to forget the Lord and trust in their own power – a perpetual temptation for God's people.

Derek Schlabach calls this the "Deuteronomic juncture [which] is the problem of how to receive and celebrate the blessing, the *shalom*, the good, or the 'land' that God desires to give, yet to do so without

defensively and violently hoarding God's blessing."[87] Tracing the history
of Christianity in a restitutionist way and trying to find a mistake or
cluster of mistakes where the early Christians began to fall into temptation neglects the basic fact that temptation and the possibility of ca-
pitulation are *always* there. Failure to understand this will cause us to
jump to conclusions regarding faithfulness and unfaithfulness. Yes,
unfaithfulness and evil may lead to further unfaithfulness, but faithfulness itself — even the good that God gives, as we have learned from
Deuteronomy — may become an occasion for temptation as well.[88]

Recall C. S. Lewis's counsel on the importance of reading old books.
In the Christian church, Lewis's advice should point us to the church
fathers, who had a consistent and common theme of commitment to
the church's sacred Scriptures. Anyone who takes the time to read a
treatise by one of them cannot miss the overwhelming presence of
Scripture that is cited, alluded to, and assumed on every page. The explicit call to the centrality of Scripture is unmistakable. For example,
in his *Homilies on Genesis*, John Chrysostom proclaims that "there is not
even a syllable or even one letter contained in Scripture which does not
have great treasure concealed in its depths."[89] He compares the study
of Scripture to people digging for metal ore:

> They don't stop short at its first appearance; instead, when they get down
> to great depth and are in a position to collect nuggets of gold, they expend
> much effort and vigor in separating them from the soil, and despite that
> great labor they find only some slight consolation for their pains. Still,
> even though they know they will gain little return in comparison with
> their trouble, in many cases despite long hours and much frustration and
> disappointment of their hopes, they don't give up at this stage: buoyed
> up by expectation they feel no effect of their efforts. So if they exhibit
> such zeal in regard to things that are corruptible and passing, to which

---

[87] Ibid., 451.

[88] Ibid.

[89] John Chrysostom, *Homilies on Genesis* 21.1 (FC 82:50). The English translations are from John
Chrysostom, *Homilies on Genesis* 18-45, trans. Robert C. Hill, FC 82 (Washington, DC: Catholic
University of America Press, 1990).

is attached much uncertainty, much more should we exhibit a like, or even greater, enthusiasm in cases where the wealth is proof against theft and the treasure is not consumed nor is it possible for hopes to be disappointed, so that we may be able to have the good fortune to enjoy the object of our zeal, reap much benefit in the process, and in the knowledge of God's ineffable love prove to be grateful to our Lord and also render ourselves immune to the devil's wiles by winning favor from above.[90]

Athanasius of Alexandria, after listing the books that belong in the Christian canon of Scripture, calls these same Scriptures "fountains of salvation, that they who thirst may be satisfied with the living words they contain."[91] In his letter to Nepotian, Jerome details the way of life and the duties of a clergyman. Paramount in these duties and way of life is to "read the Scriptures constantly; never, indeed let the sacred volume be out of your hand."[92] One need not look very far in the Fathers to read similar statements throughout their writings.

Unfortunately, it is not uncommon in evangelical circles to assume that the Fathers were not concerned with Scripture or that they let unscriptural ideas dominate their thinking.[93] In an ironic twist on this, I recall a frustrated student who was trying to write a paper on Irenaeus's understanding of apostolic succession. The student's frustration was related to the stressful time constraint of this particular assignment. She expected to turn to Irenaeus's *Against Heresies* and find a systematic and concise presentation of the theology of apostolic succession. But it took her a great deal of time because she had to wade through so much scriptural material used by Irenaeus to explain the concept. Like this student, we may have heard about the important figures of early Christianity, such as Irenaeus, who articulated the idea

---

[90]Ibid., 21.2 (FC 82:51-52).

[91]Athanasius of Alexandria, Letter 39.6 (NPNF2 4:52).

[92]Jerome, Letter 52.7 (NPNF2 6:92).

[93]Bryan Litfin identifies this as a common misconception about the Fathers. Contrary to the misconception, Litfin states, "Many of the fathers' writings are so full of scripture that you can scarcely read a single paragraph without coming across a biblical citation and allusion." See Litfin, *Getting to Know the Church Fathers*, 21.

of apostolic succession; affirmed the threefold ministry of bishop, priest, and deacon; and promoted the importance of the rule of faith.[94] But, as this student realized, these ideas are "embedded in what seem endless arguments about how to read specific biblical passages."[95]

Further, it becomes apparent when one reads Fathers like Athanasius, who battled those whom we call heretics, that the issue was not about doctrine in the strict sense but about how to properly read Scripture. No doubt, doctrine is important, but we get doctrine from Scripture, which subsequently shapes how we read other parts of Scripture. Although the Fathers adhered to the authority of Scripture, they also recognized that affirmations about scriptural authority did not necessarily lead to good interpretation. Seeing how the early church dealt with varying scriptural interpretations helps us see what was important.

In their reading and interpreting Scripture in the context of doctrine and heresy, the Fathers provide us with "exegetical guideposts out of which we dare not venture."[96] It was the church fathers who faced head-on the onslaught of objections leveled against Christianity. It was these same church fathers who met controversies and issues within the church. Their responses are classic expressions that are still important and relevant for us today. Of all the evangelicals today exhorting us to return to the Fathers, not one of them claims that we return to *everything* they discussed or taught. After all, there are significant differences even among the Fathers themselves. The call for a return is prefaced on an understanding that some things in Christian orthodoxy are essential, while others are nonessential. The Fathers help us in this important distinction.

Roger Olson gives an explanation of the difference from our own Protestant history. During the Reformation some Protestants began using the term *adiaphora* (Greek for "indifferent things") to indicate

---

[94]R. R. Reno, "The Return of the Fathers," *First Things* 167 (2006): 17.
[95]Ibid.
[96]Litfin, *Getting to Know the Church Fathers*, 28.

that some things in Christianity matter more than others.[97] In the words of Olson, "A well-balanced Christianity recognizes that some beliefs matter more than others; some truths are worth dividing over if necessary and others are not."[98] To show the differences in greater detail, he distinguishes between three categories of Christian belief: dogmas, doctrine, and opinions.[99] Dogmas are beliefs that center on the heart of Christian identity, and their denial constitutes outright heresy. Doctrines are beliefs important to a particular community of believers or denomination but are not at the very heart of Christian identity. While they are derived from Scripture, there is the recognition that there are other ways of interpreting given passages. Opinions are often speculative in nature because Scripture does not address the issues in question. As long as they do not impinge on the essentials, Christians are free to speculate on these issues.

We need not express this in the same way as Olson, but as I often tell my first-year theology classes, it is foundational to the theological task to recognize that some Christian beliefs matter more than others. Failure to make this recognition can make enemies out of friends and divide Christians over matters of relatively minor importance.

Christopher Hall offers a very good summary of the main issues that the Fathers, as leaders of the church, had to think through and resolve. The questions and answers have a continuing significance for the church today because they form the backbone of orthodoxy, or essential Christian faith. Hall writes of the responses of the Fathers as coalescing around central theological loci:

- The question of authority: To what should the church look for its guiding authority? What is the relation between Scripture and the

---

[97]For more detail on the precise context, see Justo L. González, *A History of Christian Thought*, vol. 3, *From the Protestant Reformation to the Twentieth Century* (Nashville: Abingdon, 1975), 107-11; and Carter Lindberg, *The European Reformations* (Malden, MA: Blackwell, 2001), 242-48.

[98]Roger E. Olson, *The Mosaic of Christian Belief: Twenty Centuries of Unity & Diversity* (Downers Grove, IL: InterVarsity Press, 2002), 44.

[99]Ibid., 44-45.

apostolic tradition, and how do these two relate to one another in the formation of doctrine?

- The question of the Trinity: Is Christ genuinely divine? If so, how is the divinity of Christ to be understood in relationship to the Father and the Spirit?

- The question of the incarnation: What is the relationship between Christ's deity and humanity? If Jesus was truly divine, was he also truly human? How can he simultaneously be both?

- The question of Christ's work: How has Jesus' ministry, death and resurrection overcome sin and introduced the life of the age to come into this present evil age?

- The question of humanity: What is a human being? What does the Scripture mean when it states that human beings have been created in the image of God? How and to what extent has sin affected and infected human nature?

- The question of the church: What is the church? How is the church related to Christ? What is the church's task on earth? How does one enter the church? What are the church's marks? How is the life of the church nourished and strengthened? What are the dangers the church can expect to encounter in its mission and ministry on earth?

- The question of the future: What will happen in the future? When will Christ return? What is the resurrection of the dead? What will occur at the last judgment?[100]

While the Fathers did not offer the same answers to all of these issues, they are remarkably united on the essentials, or what Ramsay calls "the rudiments of Christian confession." He states that we can confidently affirm that the Fathers agree among themselves and with us on these essentials — belief in a triune God; in a Christ who is at once

---

[100]Hall, *Learning Theology*, 18.

divine and human and who exercises a salvific role with respect to the human race; in the infallibility of Scripture; in the fallen condition of the human race and its need for salvation; in certain important rites, chief among them being baptism and Communion; in the church, in which unity must be preserved; and in the value of prayer.[101]

CONNECTION TO THE CHURCH

The context within which theology is done today is different from the context of the church fathers. Today, the vast majority of those writing theology books are located in universities, Bible schools, and seminaries, while the majority of the Fathers were pastors in the church. The context of each will inevitably affect what theologians emphasize. Today a theologian can go about his or her business without reference or preference to the church if he or she chooses. This would have been impossible in the age of the Fathers, and thus their theological reflection and biblical exegesis have a "marked pastoral emphasis and concern that is immensely practical."[102]

The interesting thing, however, is that their pastoral and practical focus did not lead them to downplay the need for deep theological reflection. The Nicene Creed (381), which expresses the foundational Christian doctrine of the relationship of the Father, Son, and Spirit (the Trinity), was forged during the age of the Fathers. The Definition of Chalcedon (451), which articulates the vital doctrine of the relationship of the human and the divine in Jesus Christ, was also formed during this time. It is important to recognize that these statements of essentials came about as the leaders of the church wrestled with these difficult theological issues. Today we tend to assume that these principal doctrines are clear and plain to see in the Bible. But what we fail to understand is that these classic formulations took many years to work out as church leaders wrestled with the biblical texts. This took place

---

[101]Ramsay, *Beginning to Read*, 6.
[102]Hall, *Reading Scripture*, 54.

in the context of the church because the Fathers saw it as vitally important to the spiritual well-being of all Christians.

The location of theological reflection in the church meant that the Fathers had it in the forefront of their theologizing. For us, this communicates two main things. First, the Fathers show that deep theological thought and reflection are not antithetical to a deep spiritual life – in fact they are required. Second, theological study done for the church has significance for Christian believers. The Fathers show us that the theologian can "blend profundity and practicality" because as pastors their ultimate concern was for the spiritual well-being of their congregations.[103] We should not, therefore, let the perceived remoteness of the theological discussion done by the Fathers keep us from recognizing their vital importance to the church.

If we take the ecumenical (universal) councils as examples of the connection of the church (and therefore the Fathers) to theology, we see that they were inspired by the very practical issue of human salvation.[104] The central message of the Christian faith is that humanity is separated from God by sin. Solo efforts to break the wall of separation are impossible, so the initiative is taken by God to remedy the situation. God becomes man, is crucified, and rises from the dead, thereby redeeming sinful humanity from its bondage. This is the foundational message that the church councils and creeds sought to safeguard and hand on. The reason heresies were taken so seriously and required condemnation at these councils was because they undermined this central message of the Scriptures and, in doing so, erected a new barrier between humanity and God. In other words, they undermined the very message of salvation.

The councils addressed foundational issues pertaining to our salvation. No one less than God can save humanity, so if Christ is to save,

---

[103]Ibid.

[104]What follows draws on Timothy Ware, *The Orthodox Church*, new ed. (London, UK: Penguin, 1997), 20-22.

he must be God. But in order for humanity to participate in God's redemption for us through Christ, he must be truly human as we are. Each heresy confronted by the councils addressed some aspect of this foundational affirmation. According to Timothy Ware,

> The first two, held in the fourth century, concentrated upon the earlier part (that Christ must be fully God) and formulated the doctrine of the Trinity. The next four, during the fifth, sixth, and seventh centuries, turned to the second part (the fullness of Christ's humanity) and also sought to explain how humanity and Godhead could be united in a single person. The seventh council, in defence of the Holy Icons, seems at first to stand somewhat apart, but like the first six, it was ultimately concerned with the Incarnation and with human salvation.[105]

But it is not only through the councils that the Fathers' profundity and practicality can be seen. They also show us how to unite heart and mind in theology.[106] In this they have something to teach those of us in the evangelical tradition who might tend toward an understanding of theology that is overly rationalistic. The Fathers show us that mere intellectual assent to a list of doctrines is an impoverished Christianity and one that needs correction and supplement.

A good example is found in Gregory of Nazianzus's (329–390) emphasis on mystery as a correction to an overemphasis on reason. Gregory argues in his *First Theological Oration* that those whose focus on theological propositions to which one should or should not intellectually assent run the risk of making "our Great Mystery . . . a thing of little moment."[107] Theology, he states, can go to excess when we go beyond our intellectual bounds and assume we understand the mystery. Reason has its place, but it also has its limits. This is why the theologian should be affected by the subject of theology. Gregory thus exhorts theologians to "look to ourselves, and polish our theological self to beauty like a

---

[105]Ibid., 23.
[106]Ramsay, *Beginning to Read*, 19.
[107]Gregory of Nazianzus, *Oration* 27.2 (NPNF[2] 7:285).

statue."[108] This is done through things like commending hospitality, showing love, feeding the poor, singing psalms, fasting, and praying, and through mastering our passions, pride, greed, thoughts, and the like.[109]

Gregory explicitly builds on this line of thought in his *Second Theological Oration*.[110] The character of theologians matters as they enter the theological task. But it does not end there. In a beautiful section Gregory compares the theological task with the Exodus account of Moses on Mount Sinai. With the proper character, one may "go up eagerly into the Mount."[111] But this approach to "hold converse with God" is done simultaneously with longing and fear because of the limitations inherent in our material existence and humanity. This is why, on Sinai, Moses was not permitted to see God's face, only his back.[112] God's essence or being is signified for Gregory by his face, while God's actions or operations are signified by his back:

> For these are the Back Parts of God, which He leaves behind Him, as tokens of Himself like the shadows and reflection of the sun in the water, which shew the sun to our weak eyes, because we cannot look at the sun himself, for by his unmixed light he is too strong for our power of perception. In this way then shalt thou discourse of God.[113]

The Fathers' recognition that our talk about the divine is limited by our finitude and God's eternality gives them a view of theology that we would do well to consider and even seek to emulate. I pointed out above that Gregory of Nazianzus believes an overconfidence in our ability to reason about God trivializes "our Great Mystery." This may sound counterintuitive to those who have read the Fathers, especially given their penchant for deep theological thought. But their theological task was always understood within the context of mystery and qualified by humanity's limits.

---

[108]Ibid., 27.7 (NPNF² 7:287).
[109]These are all expressed negatively in the *First Theological Oration*.
[110]Gregory of Nazianzus, *Oration* 28 (NPNF² 7:288-301).
[111]Ibid., 28.2 (NPNF² 7:289).
[112]Ex 33:23.
[113]Gregory of Nazianzus, *Oration* 28 (NPNF² 7:289).

Many people could probably point to the first book that sparked their love for reading. For me it was an Agatha Christie book assigned in my grade seven year called *And Then There Were None.*[114] Published in 1939, it is considered Christie's masterpiece in the genre of mystery. In the novel, ten people are enticed to go to an island under varying pretexts, whether it be an offer of employment or to reacquaint with old friends. All ten have, in some way, been complicit in the deaths of other people but have escaped punishment for their crimes. After the dinner on the first night of their arrival all the guests are charged with their "crimes" through a gramophone recording and informed that they have been brought to the island to pay for their crimes. Escape is impossible due to the island's distance from the mainland and inclement weather. They are the only people on the island, and one by one each is killed in a manner parallel to ten deaths in a nursery rhyme called "Ten Little Indians." After the last death, nobody else appears to be left on the island. But in a postscript at the end of the novel, a confession reveals who was responsible for the killings and how they took place. As someone reads the novel, he or she plays the part of detective, investigating the mystery in order to solve it. This is what captured my imagination as a child. My task was set – solve the mystery and find the killer!

Today when we speak of mystery, what we usually mean is something like what I have just described. Our tendency is to think of a mystery as something that is solved through careful investigation. We could call this "investigative mystery." But this was not the kind of mystery Gregory or other church fathers had in mind. Mystery, for them, "does not so much confront me, as envelop me, draw me into itself; it is not a temporary barrier, but a permanent focus of my attention."[115] In this approach to mystery we don't seek to solve it as much as we seek to participate in it. The Fathers did turn to Scripture

---

[114]The book has also been published under the title *Ten Little Indians.*
[115]Andrew Louth, *Discerning the Mystery: An Essay on the Nature of Theology* (Oxford: Clarendon Press, 2003), 68.

to "know" God, but this knowledge was not discerned through an "investigative mystery" approach. As we saw in Gregory of Nazianzus's appeal to Moses on Sinai as an analogy to our approach to God, God is not a problem to be investigated and solved as a result of investigation. For the Fathers, theology is not a matter of solving a mystery but of participating in it.

Steven Boyer and Christopher Hall call this "revelational mystery" and argue that it is already well established in the New Testament.[116] In Mark 4:11 Jesus describes the apostles as those who have "been given the [mystery] of the kingdom of God."[117] Jesus was not saying that he had given the apostles the key to a puzzle that needed solving. The purpose of the mystery is not for the apostles to be "in the know" when the secret is given. If this were the case, the mystery would no longer be mysterious to them. We know this is not the case simply because of the times in the Gospels when the apostles express confusion about some of the things that occur.

Similarly, Paul sought to make known "the mystery that has been hidden throughout the ages and generations but has now been revealed to his saints."[118] He insists that "the mystery was made known to me by revelation,"[119] but still confesses, "O the depth of the riches and wisdom and knowledge of God! How unsearchable are his judgments and how inscrutable his ways!"[120] For Paul, the plan of God is not an investigative mystery that, once communicated, is solved. Rather, it is communicated *as* a mystery. This is how the term *mystery* is used throughout the New Testament:

- The hardening of Israel is a mystery (Rom 11:25).

- The final resurrection of the dead is a mystery (1 Cor 15:51).

---

[116]What follows is dependent on Steven D. Boyer and Christopher A. Hall, *The Mystery of God: Theology for Knowing the Unknowable* (Grand Rapids: Baker Academic, 2012), 5-7.

[117]Unfortunately, the NRSV translates *mystērion* as secret. I have taken the liberty to insert "mystery" as a better translation.

[118]Col 1:26.

[119]Eph 3:3.

[120]Rom 11:33.

- The summing up of all things in Christ is a mystery (Eph 1:9-10).

- The inclusion of the Gentiles in the church is a mystery (Eph 3:9).

- The union of husband and wife as a picture of Christ and the church is a mystery (Eph 5:31-32).

- "Christ in you, the hope of glory" is a mystery (Col 1:27).

- "Christ himself, in whom are hidden all the treasures of wisdom and knowledge" is "God's mystery" (Col 2:2-3).

In these examples the link of mystery to revelation – its being made known – is clear, but even here the mystery does not cease being a mystery. This is why a "revelational mystery" is different from an "investigative mystery." An investigative mystery seeks a solution, but a revelational mystery remains a mystery even after it is revealed. In fact, "It is precisely in its revelation that its distinctive character as mystery is displayed."[121]

Mystery as problem solving (investigative) is concerned with what is known and able to be grasped. But revelational mystery revolves around what is unknown and ungraspable – this is why it always remains mystery even though it is revealed. Because of the Fathers' insistence on revelational mystery, they can show us where we may just miss the point in theological study. If the purpose of proper Bible study is verifiable data, then the proper function of theology is the systematic organization of that data with which we can speak with certainty to the world.[122] The Bible would then be treated merely as a source book of information for theology and other things. Once the information is mined from the Bible we would then have our system, and the mystery is solved. Not only does this run the risk of making the Bible superfluous, it seeks to remove "our Great mystery," and we may just think we have "solved" God. Then he must fit into our categories and he

---

[121]Boyer and Hall, *Mystery of God*, 6.
[122]Clark Pinnock, "Theological Method," in *New Dimensions in Evangelical Thought: Essays in Honor of Millard J. Erickson*, ed. David D. Dockery (Downers Grove, IL: InterVarsity Press, 1998), 201-2.

becomes the God we think he should be. The Fathers encourage us to let God be God.

In this chapter I have attempted to communicate the importance of these early leaders of the Christian church for believers today. Truly understanding this may take time and deliberation. In a culture of instant gratification, it will require discipline not to think of them as another quick fix to the things that ail the church or as icing on top of the creation science cake. But this is my point in commending the Fathers to evangelicals today. The Fathers offer us a window into an age where some of the most foundational Christian ideas, doctrines, and practices were discovered and developed. The cure for our amnesia is not ignoring them.

RECOMMENDED READING

"The First Bible Teachers: Reading over the Shoulders of the Church's Founding Fathers." *Christian History Magazine* 80, no. 4 (2003).

Hall, Christopher A. *Reading Scripture with the Fathers*. Downers Grove, IL: InterVarsity Press, 1998.

——— . *Learning Theology with the Fathers*. Downers Grove, IL: InterVarsity Press, 2002.

Litfin, Bryan. *Getting to Know the Church Fathers: An Evangelical Introduction*. Grand Rapids: Brazos Press, 2007.

Wilken, Robert L. *The Spirit of Early Christian Thought: Seeking the Face of God*. New Haven, CT: Yale University Press, 2003.

2

HOW *NOT to* READ *the* FATHERS

A SURVEY OF CREATION SCIENCE APPROPRIATION OF THE FATHERS

At the risk of sounding alarmist, I begin this chapter by relating its topic to a fictitious scenario that Alasdair MacIntyre calls "a disquieting suggestion."[1] MacIntyre asks his readers to imagine a scenario where the general public blames a series of environmental disasters on scientists. Scientists are killed and there is widespread destruction of laboratories, books, and instruments. A grassroots political movement gains power and successfully lobbies to abolish science teaching in schools and universities, and imprisons or kills what scientists remain. After a significant amount of time, a movement of people reacting to the new state of affairs arises. Enough time has passed that now there is a significant gap in scientific knowledge and especially in understanding science itself. This is the context in which the group tries to reestablish the scientific method. The problem is that they have only disjointed concepts on which to rebuild. The original theoretical context

---

[1]Alasdair MacIntyre, *After Virtue: A Study on Moral Theory* (Notre Dame, IN: University of Notre Dame Press, 1981), 1-2.

does not come into consideration, and this disconnect drastically affects their incomplete understanding of experiments. Proper use of instruments that survived the destruction has been forgotten. Still, even in this fragmentary and incomplete state, these are all re-embodied in a series of practices that go by the name of science.

In this state, arguments occur over theories even though they are presented incomplete. In schools children memorize things like the periodic table and recite certain theorems even though they are detached from the context in which they have meaning and significance. Everyone thinks they are doing science, but it is not really science at all. People use scientific words and language, but "many of the beliefs presupposed by the use of these expressions would have been lost and there would appear to be an element of arbitrariness and even of choice in their application which would appear surprising to us."[2]

There is a parallel between this fictitious account and the way the church fathers are sometimes used. As I discussed in the previous chapter, our historical and theological connections to the early church have been severed by a particular reading of Christian history. Recently that reading has been called into question, and a rediscovery of patristics is underway in some quarters. But it is sometimes done in the way MacIntyre describes his story of science – with "an element of arbitrariness and even of choice." The use of theories and instruments was being undertaken devoid of meaning and context and at the whim of the "scientist." The same could be said about the way some employ the church fathers, and in this chapter we will examine how some use the early church leaders in the creation/evolution debate.

In the 1985 edition of Boniface Ramsay's *Beginning to Read the Fathers*, Ramsay relates a description about popular knowledge of the Fathers in Christianity, most of which I believe holds true today, particularly in evangelicalism:

---

[2]Ibid., 1.

The Fathers have always been prominent in the life of the Church, and they have constantly been invoked for every sort of thing. But they have suffered from their prominence; they are famous but not well known. Like so many famous people, they are cited and alluded to, but few go to the effort of exploring their thought.[3]

In disagreement, one might point to the abovementioned and well-documented attention many in the evangelical orbit have given to these early Christian leaders. This is undoubtedly true – at a certain level evangelicals have, in fact, been rediscovering the church fathers and their context. For example, one can see the concerted effort of Christian publishers to expose the broader evangelical audience to the Fathers.[4]

One wonders, however, what kind of impact this is actually having in the broader evangelical world. Bryan Litfin, professor of theology at Moody Bible Institute, addresses this in his excellent introduction to the church fathers, intended specifically for an evangelical audience. He chastises those who use the Fathers merely as ammunition in a debate. "Too often," he laments, "a snippet from an ancient writer is yanked out of context to support a modern viewpoint. Such an approach is unfair to authors who never intended that their writings be excerpted out of their whole corpus to serve as ammunition in a modern-day war of words."[5] This misappropriation likely spurred D. H. Williams to caution that we not "create the fathers in our own image."[6]

---

[3]Boniface Ramsay, *Beginning to Read the Fathers* (Mahwah, NJ: Paulist Press, 1985), 3.

[4]See, for example, Jason Byassee, *Praise Seeking Understanding: Reading the Psalms with Augustine* (Grand Rapids: Eerdmans, 2007); Hans Boersma, *Heavenly Participation: The Weaving of a Sacramental Tapestry* (Grand Rapids: Eerdmans, 2011); Christopher A. Hall, *Reading Scripture with the Fathers* (Downers Grove, IL: InterVarsity Press, 1998); Christopher A. Hall, *Learning Theology with the Fathers* (Downers Grove, IL: InterVarsity Press, 2002); Bryan Litfin, *Getting to Know the Church Fathers: An Evangelical Introduction* (Grand Rapids: Brazos Press, 2007); Mark Sheridan, *Language for God in the Patristic Tradition: Wrestling with Biblical Anthropomorphism* (Downers Grove, IL: IVP Academic, 2015); Robert E. Webber, *Ancient-Future Faith: Rethinking Evangelicalism for a Postmodern World* (Grand Rapids: Baker, 1999); D. H. Williams, *Evangelicals and Tradition: The Formative Influence of the Early Church*, Evangelical *Ressourcement* Series (Grand Rapids: Baker Academic, 2005).

[5]Litfin, *Getting to Know the Church Fathers*, 14-15, 28.

[6]D. H. Williams, "*Similis et Dissimilis*: Gauging Our Expectations of the Early Fathers," in *Ancient Faith for the Church's Future*, ed. Mark Husbands and Jeffrey P. Greenman (Downers Grove, IL: IVP Academic, 2008), 70.

As an important voice in the evangelical appropriation of the Fathers, Williams welcomes the recent attention that has been directed toward the early church. But his years of studying patristics have also made him "cautious about this newfound enthusiasm in the early fathers for the reason that evangelicals might be tempted to tame the early fathers by making them speak to our current situation in ways alien to the ancients themselves."[7]

Williams warns evangelicals against picturing the church fathers as having a monolithic stance on doctrine and practice. There were some major differences centering on significant theological doctrines. For example, Athanasius battled Arianism his whole life. While it is true that Arianism was eventually judged heretical because of its denial of the deity of Christ, one should not miss the important point that both Arius and Athanasius "were drawing on selected portions of a body of tradition that existed and yet was not formulated to answer the new questions of their time that were being raised."[8] Arius was a presbyter in Alexandria – he was not bent on corrupting the faith. Rather, he was offering an interpretation of Scripture that was ultimately judged as heretical.

But it was not only a matter of orthodox versus heretical teaching that showed differences in practice and doctrine. Sometimes the faithful conflicted with the views of other Christians. The most famous example of this is the christological controversy of the fifth century.[9] All the church leaders involved agreed on the divinity of Christ and sought to guard that divinity. But they were radically opposed to each other's logic and terminology, which sought to express that divinity in light of the incarnation.

According to Williams, the danger of using the age of the Fathers as a wondrously united age is that

---

[7]Ibid.

[8]Ibid., 78.

[9]An accessible introduction to the issue in its context is Stephen W. Need, *Truly Divine and Truly Human: The Story of Christ and the Seven Ecumenical Councils* (Peabody, MA: Hendrickson, 2008).

early Christianity becomes too easily subject to our own agendas for using it. It is difficult to disprove general claims that make appeal to a church age that never existed – the ancient church becomes another Lake Wobegon, "where all the women were strong, the men good-looking and the children above average." . . . We do this at the cost of misconstruing ancient historical and theological realities. Before we are tempted to remake the patristic church in our own image, let us first encounter it with its own problems and solutions.[10]

Williams argues that this is a "potentially serious problem" for evangelicals who are using the Fathers. The problem is exacerbated by merely presenting evangelical readers with bite-size portions from their writings, yanked, as Litfin bemoans, out of the context and flow of argument.

Williams's article was written, in part, to address those who inappropriately use the Fathers to "look for proto-Protestant doctrines or practices in the ancients that support the Evangelical view."[11] The problem for Williams is that these concerns are polemical – they are reading the ancients through a post-Reformation Protestant lens and looking for Protestant doctrines in the early church so that they can claim vindication for their beliefs and subsequently claim to be upholders of the true faith. For Williams, this approach is "tantamount to approaching their writings as a theological grab bag, hoping to find some foundational principle that reinforced the claims of one church against another."[12]

I see a parallel to Williams's concerns of inappropriate use of the Fathers in Protestant attacks on Catholicism in the way the Fathers are employed in the creation/evolution debate. One glaring example is the very title of Louis Lavallee's essay "The Early Church Defended Creation

---

[10]Williams, "*Similis et Dissimilis*," 79.

[11]Ibid., 80. Williams states, "There have been recent attempts to find a 'patristic principle of *sola scriptura*' in Irenaeus or in Athanasius, from which the conclusion is reached, '*Sola scriptura* has long been the rule of believing Christian people, even before it became necessary to use the specific terminology against later innovators who would usurp the Scriptures' supremacy in the church'" (ibid.); quoting James White, "Sola Scriptura and the Early Church," in *Sola Scriptura*, ed. Don Kistler (Morgan, PA: Soli Deo Gloria, 1997), 53.

[12]Williams, "*Similis et Dissimilis*" 80.

Science."[13] Here we see what may be called a proto–creation science appeal to the Fathers similar to the proto-Protestant appeal criticized by Williams. How could the Fathers defend something that was unheard of in the patristic age? I echo the concerns of Litfin and Williams, and it is in this spirit that I wish to point out several manifestations of these concerns.

The church fathers have indeed been marshaled as evidence and support from all sides in the creation/evolution debate. For example, old-earth creationist Hugh Ross asks his readers to look to the "wisdom of the ages" to inform our understanding of the days in Genesis 1.[14] Some "young-earth creationist leaders," he explains, have attempted to look into the Christian past and claim that a twenty-four-hour creation day interpretation was the norm until the advent of modern geology and Darwinism. Thus, Ligon Duncan and David Hall claim that the day-age view is only as old as the rise of jazz music in America.[15] According to Duncan and Hall, the twenty-four-hour-day view has been the consensus position of the church for a very long time: "If ever the Church agreed on anything, it has been on the days of creation."[16] It was recent naturalistic/rationalistic paradigms that forced Scripture to occupy a subservient role.

---

[13]Louis Lavallee, "The Early Church Defended Creation Science," accessed September 30, 2017, www.icr.org/article/early-church-defended-creation-science/. Accessing the print or PDF version of this article has proven frustrating. At the end of the web page the reader is directed to cite the article as from *Acts & Facts* 15, no. 10. *Acts & Facts* is the journal for the Institute for Creation Research. Back issues for the journal can be found at www.icr.org/index.php?module=articles&action=search&f_typeID=7. I searched every issue for the year 2015 and could not find this article.

[14]Hugh Ross, *A Matter of Days: Resolving a Creation Controversy* (Colorado Springs: NavPress, 2004), 41-49. "Wisdom of the Ages" is the title of chapter four. Unless otherwise indicated, explanation of Ross's position will be drawn from this chapter.

[15]"The day-age theory takes the 'days' in Genesis 1 as periods of indefinite length, such that neither the age of the earth nor the duration of any particular period in creation history can be determined from the Bible" (Ted Davis, "Science and the Bible: Concordism, Part 1," June 19, 2012, http://biologos.org/blogs/ted-davis-reading-the-book-of-nature/science-and-the-bible-concordism-part-1#sthash.5xRLZEup.dpuf).

[16]The twenty-four-hour view holds that the creation days of Genesis indicate six consecutive twenty-four-hour periods. See J. Ligon Duncan III and David W. Hall, "The 24-Hour View" and "The 24-Hour Reply," in *The Genesis Debate: Three Views on the Days of Creation*, ed. David G. Hagopian (Mission Viejo, CA: Crux, 2001), 47, 52, 99.

For Ross, a "look at what actually occurred in the past" should help clarify Duncan and Hall's misconception.[17] Rather than overwhelming agreement on the length of days in Genesis, he sees diverse opinions in both the Jewish and Christian traditions, and this diversity is significant. He recognizes that many Protestant traditions downplay the significance of "past church scholars," but he argues that the "historical position of the church carries great authority where we find clear unanimity, in matters such as the deity of Jesus Christ and the means of redemption from sin."[18] Thus, this diversity indicates for Ross a hint about the relative importance of the length of days in the early church.

In Ross's brief proof-texting of various church fathers and their understanding of the days of Genesis he does not, however, demonstrate this diversity.[19] His citations of select church fathers all lean toward a nonliteral reading of the days, along with certain Fathers who represent an ambiguity when it comes to the days in Genesis. Ross does not really show the breadth of diversity he claims exists in the Fathers, but he does cast a shadow of doubt on the absolute claim of authors like Duncan and Hall who espouse a twenty-four-hour view.

In "The Church Fathers on Genesis, the Flood, and the Age of the Earth," James Mook takes Hugh Ross to task for using the church fathers against six-day creationism and the twenty-four-hour-day view.[20] Ross's claim, according to Mook, that the Fathers were interested in *theological* rather than *historical* meaning in Genesis is not well-supported by the Fathers themselves. For Mook, it is inaccurate to suggest that Irenaeus,

---

[17]Ross, *Matter of Days*, 41.

[18]Ibid., 41-42.

[19]Ross observes passages from the following church fathers: Justin Martyr, Irenaeus, Hippolytus, Clement of Alexandria, Origen, Lactantius, Victorinus of Pettau, Methodius of Olympus, Augustine, Eusebius, Basil, and Ambrose.

[20]James R. Mook, "The Church Fathers on Genesis, the Flood, and the Age of the Earth," in *Coming to Grips with Genesis: Biblical Authority and the Age of the Earth*, ed. Terry Mortenson and Thane H. Ury (Green Forest, AZ: Master Books, 2008), 25.

Origen, Basil, and Augustine were "day-age proponents."[21] So Mook sets out to demonstrate the inaccuracy of Ross's claims about the Fathers:

> A natural reading of the Church fathers shows that though they held diverse views on the days of creation, and correctly gave priority to the *theological* meaning of the creation, they definitely asserted that the earth was created suddenly and in less than 6,000 years before their time. They left no room for the "old earth" views promoted by Ross and other moderns.[22]

To ask who has the correct understanding of the Fathers on this issue would be to miss the point I am trying to make. This is precisely where the concerns of Litfin and Williams become relevant. Both Ross and Mook marshal "evidence" to support their particular position, but we should consider how this evidence is handled.

It is significant that Mook is also concerned with mishandling of the church fathers. In the section of his essay titled "The Importance of the Church Fathers to the Age Controversy," he offers an apologetic for the importance of the Fathers, who have been vitally important in "clarifying the parameters of orthodoxy."[23] He recognizes the renewal of interest in the Fathers, particularly in how they handled matters like the length of the days of creation and the age of the earth. But in spite of Mook's appreciation for the importance and use of the Fathers, he warns that there is a "tendency with some to misread the patristic literature."[24] These misreadings are similar to what can occur in our reading of the Bible: failure to consider context, eisegesis, and the teachings of the Fathers being "muffled altogether."[25]

For Mook, too many scholars misrepresent the Fathers' views on things pertaining to creation and evolution, leading to a troubling

---

[21]Ibid.

[22]Ibid. Italics in original.

[23]Ibid., 24.

[24]Ibid.

[25]Ibid. I must admit a difficulty in discerning exactly what Mook means here. I take him to mean something similar to what Williams wrote regarding the problems with understanding the Fathers as having a monolithic position on doctrine and practices. Thus, I consider the "muffling" to occur because the significance of each individual Father gets lost in overgeneralizations.

diversity of opinion that only serves to confuse the layperson. He therefore offers four important questions he insists should be posed to those who employ the Fathers in the creation/evolution debate:

> First, which specific ancient treatises were these modern scholars using to class the ancients into such post-Darwinian sounding categories? Second, were there any treatises or resources these modern writers over-looked? Third, if there were overlooked resources, was this innocent oversight due to perhaps consulting only secondary sources? And fourth, if these men were presented with sufficient patrological counter-evidence, would they acknowledge this in subsequent writings?[26]

Mook's explicit purpose in the entire article is to counter certain mis-readings of the Fathers by asking these kinds of questions. He thus argues for some sort of guidance in how we read the Fathers in the creation/evolution debate, and I concur with this call for guidance. Unfortunately, it is questionable whether Mook follows his own advice. In what follows I will explain some of the ways in which the Fathers are used by the creation science side of the creation/evolution debate. Collectively, these examples amount to a demonstration of how *not* to use the Fathers.

GENERALIZATION AND EISEGESIS
OF THE CHURCH FATHERS

In an Answers in Genesis (AiG) article seeking to show that the church fathers interpreted Genesis literally, the anonymous author states: "Some accuse creationists of taking an excessively literal view of Scripture, particularly Genesis, and that this view is something of recent origin, that Christians in the early Church took a more alle-gorical view of things. The writings of Basil show this not to be true."[27]

---

[26]Ibid.

[27]Anonymous, "Genesis Means What It Says: Basil (AD 329–379)," accessed September 30, 2017, http://creation.com/genesis-means-what-it-says-basil-ad-329-379. Identification of this article's author is elusive. According to the first footnote in the article, it was adapted from David Watson, "An Early View of Genesis One," *Creation Research Society Quarterly* 27, no. 4 (1991): 138-39. The anonymous version also appears at https://answersingenesis.org/genesis/genesis-means-what-it-says-basil-ad-329-379/.

But we must ask, even if Basil was a literalist,[28] would this show that the entire early church was also? It is a misrepresentation of the Fathers to claim that because one was a literalist, all others were as well. Anyone with even a scant knowledge of Origen of Alexandria would offer him as a glaring exception, but he is not the only one. In fact, the conclusion that this anonymous writer is trying to disprove is much closer to the truth than what he is trying to prove. There is a problem with both his generalization and his conclusion.

In the same vein of interpretation, James R. Mook is also guilty of generalization in his treatment of biblical interpretation in the early church. Mook divides the church fathers into "literalists" and "allegorists."[29] The literalists are said to agree with him that the creation days were twenty-four-hours long, while the allegorists do not. However, his distinction between literal and allegorical is made with absolutely no discussion about what these terms mean. And since the conservative Christian audience likely to read Mook's article is already inclined to think poorly of allegory, it is also predisposed to accept his distinction. He asserts that there existed a "tension between allegorists and literal interpreters" in the early church – again, with no attempt to define the terms or explain the context.[30] He then simply lists certain followers of each "method" and proof-texts them for support in his argument.

Several assumptions are at work in Mook's article, which are shared by his audience. For someone who claims to care about the patristic hermeneutical context, this demonstrates a troubling misuse of the Fathers. Mook assumes that allegorical interpretation is wrong and literal interpretation is right. But surely he should discuss the meaning of these terms in the context of the early church. Mook simply leaves the readers to connect the dots, which are that the "literalist" church fathers were,

---

[28]I argue below, in chapter four, that this is not an accurate portrayal of Basil's interpretation of Genesis 1.

[29]Mook, "Church Fathers on Genesis," 29.

[30]Ibid.

in fact, evangelical in their approach to Scripture. After all, don't evangelicals practice literal interpretation too? The assumption that the literalists won out is a gross misrepresentation of the context of early Christianity that is only further misrepresented by Mook's failure to define his terms.[31]

Misrepresentation of the hermeneutical context of early Christianity is commonplace in creation science material dealing with the Fathers. It leads to a host of other problems that also distort what concerned the Fathers in their understandings of Genesis. For example, as indicated above, some posit Basil of Caesarea as a literalist because of his comments about allegory in Homily 9 of his *Hexaemeron*. But this labeling of Basil fails to understand not only his context but also the broader hermeneutical context of early Christianity. This is precisely why I have devoted an entire chapter to the hermeneutical context of the early church, especially as it relates to the apparent divide between the literalists and allegorists. Further, I have devoted a chapter to the question of whether Basil was, in fact, a literalist.

Mook's treatment of Origen of Alexandria shows how his generalizations and failures to define terms lead him to misrepresent his subject. He recognizes that Origen was "one of the greatest minds in Christian antiquity" but qualifies this with the claim that "his teachings are now recognized as aberrant in significant ways."[32] Mook identifies Origen as "one of the main formulators of the allegorical hermeneutic in the early Church."[33] Already Mook has cooked the books against Origen. He appears in the article as the chief allegorist, and probably one of the worst kind since his teachings (all of them?) are significantly aberrant. So, to an audience that is already predisposed against allegory (even though they have not been informed about what it actually is), it has also been linked to aberrant teaching.

---

[31]This context is discussed in detail in chapter three.
[32]Mook, "Church Fathers on Genesis," 33. Mook repeats this two times on page 34 and once on page 35.
[33]Ibid.

Surely Mook's comments about Origen beg for qualification. It is true that Origen was, and remains, a controversial figure in Christianity. But there are too many sweeping generalizations about Origen and allegory here that obscure exactly *why* he was, and is, controversial. Patristics scholar John A. McGuckin writes this about Origen:

> The most influential of all Greek theologians, Origen was the architect (whether his opinions were followed or were explicitly rejected) of most of the substructure of Christian dogma and biblical theology in the late antique period of Christianity. His influence was as great as that of Augustine in the West, although in the Greek-speaking world the variety of other major thinkers moderating and redirecting the channels of his thought (such as Gregory of Nazianzus and Maximus the Confessor) ensured that his intellectual legacy would be more creatively received and developed.[34]

What are we to make of this description of Origen as one of the most influential scholars in the early church? It is true that one can be influential and still teach unorthodox or heretical doctrines – just ask Arius. But Origen is not necessarily described in this way by McGuckin, who argues that some followed Origen while others did not. Mook fails to make this acknowledgment, perhaps because it does not suit his purpose in throwing allegory on the rubbish heap of interpretive "methods" along with other apparently aberrant teachings of Origen.

One does not even have to go to scholarly treatises and articles to correct Mook's misrepresentation of Origen. Even the sometimes maligned Wikipedia reveals that not all of Origen's teachings were aberrant: "His teachings on the pre-existence of souls, the final reconciliation of all creatures, including perhaps even the devil (the *apokatastasis*), and the subordination of God the Son to God the Father, were rejected by Christian orthodoxy."[35] Mook fails to explain that, even though

---

[34]John A. McGuckin, "Origen of Alexandria," in *The Westminster Handbook to Patristic Theology*, ed. John A. McGuckin (Louisville, KY: Westminster John Knox, 2004), 243.

[35]Wikipedia, s.v. "Origen," accessed November 2, 2016, https://en.wikipedia.org/wiki/Origen#cite_note-74.

there was some controversy surrounding Origen's hermeneutical approach, it was never condemned by Christian orthodoxy and was followed, in some form, by the vast majority of church fathers after him. Even Basil of Caesarea, whom Mook labels a literalist who condemns allegory, owes a great debt to Origen and follows him in much of his interpretational approach.[36]

At times, Mook uses the classification of certain Fathers as allegorists to discredit their teachings on other things. For example, under the heading "The Allegorists," Clement of Alexandria (ca. 150–ca. 215) is identified as teaching that "God created everything 'at once' and 'together.'"[37] Mook explains that this would later be espoused by fellow allegorists Origen and Augustine, implying that allegory led them to this conclusion. But what he fails to indicate or realize is that this view was also espoused by Basil, whom Mook champions as a literalist.

The view under consideration, which has come to be called *simultaneous creation*, was a response to a theological dilemma the Fathers faced. Paul Blowers affirms that the first words of the Bible, "In the beginning," are "at the root of a gradually developing Christian theology of the beginning of the world."[38] For the most part, early patristic interpretation of this phase is unconcerned with the temporal aspects of the phrase because they believed that it had a far richer meaning. This is why the Fathers connected it to John 1:1 and Proverbs 8:22 and identified the beginning as the Logos, Jesus.[39] So, generally speaking, "Christian exegetes from all around understood *archē* [beginning] in Gen 1:1 and Jn 1:1

---

[36]See below, chapters three and four.

[37]Mook, "Church Fathers on Genesis," 33.

[38]Paul Blowers, *Drama of the Divine Economy: Creator and Creation in Early Christian Theology and Piety*, Oxford Early Christian Studies (Oxford: Oxford University Press, 2012), 140.

[39]Jn 1:1: "In the beginning was the Word." Prov 8:22: "The LORD created me [Wisdom] at the beginning of his work." This connection was made by Theophilus of Antioch, *Ad Autolycum* 2.10; Tatian, *Address to the Greeks* 5; Tertullian, *Against Hermogenes* 20.1-4; Clement of Alexandria, *The Stromata* 6.7.58; Origen, *Homily on Genesis* 1.1; Origen, *Commentary on John* 1.17.104; Basil, *Hexaemeron* 1.6; Jerome, *Hebrew Questions in Genesis* 1.1; Ambrose, *Hexaemeron* 1.4.15; and Augustine, *Unfinished Literal Commentary on Genesis* 3.6.

to have a distinctly non-temporal, ontological meaning."[40] Most exegetes, in fact, attempted to usurp philosophical traditions that used the term as "principle" or "causal origin" rather than beginning in time.

Still, in his first homily on the Hexaemeron, Basil preaches that even the meaning of *beginning* as temporal has validity. But this presented an acute problem for early Christian theologians. Blowers asks, "How could the timeless Creator produce a time-bound world without compromising his transcendence? How could eternity and time conceivably overlap in the 'moment' God created?"[41] Basil argues that the act of creation is "instantaneous and timeless" and that "beginning is something immeasurable and indivisible."[42] The actual beginning of the world is not time, because time did not exist before the creation of the world. The beginning is not part of time; it initiates what follows.[43] Thus, there was a "condition older than the birth of the world, without beginning or end" where God perfected his works.[44] Creation existed already in the mind of God; it is a "pretemporal, conceptual creation in God."[45] The unfolding of this creation is what Genesis actually relates. If Mook can question how others use the church fathers, surely we can ask him if he follows his own advice – "If these men were presented with sufficient patrological counter-evidence, would they acknowledge this in subsequent writings?"[46]

Louis Lavallee is also guilty of generalizing the Fathers. In order to show that the early church taught that God created out of nothing, he appeals to Theophilus of Antioch (d. ca. 183–185), who wrote an apology

---

[40]Blowers, *Drama of the Divine Economy*, 142.

[41]Ibid.

[42]Basil, *Hexaemeron* 1.6 (FC 46:11). All quotations of *Hexaemeron* are from *Saint Basil: Exegetic Homilies*, trans. Agnes Clare Way, FC 46 (Washington, DC: Catholic University of America Press, 1963).

[43]Andrew Louth, "The Six Days of Creation According to the Greek Fathers," in *Reading Genesis After Darwin*, ed. Stephen C. Barton and David Wilkinson (Oxford: Oxford University Press, 2009), 48-49.

[44]Basil, *Hexaemeron* 1.5 (FC 46:9).

[45]Peter C. Bouteneff, *Beginnings: Ancient Christian Readings of the Biblical Creation Narratives* (Grand Rapids: Baker Academic, 2008), 133.

[46]Mook, "Church Fathers on Genesis," 24.

called *To Autolycus*, which, according to Lavallee, "contained an extensive treatment of creation and became a model for other fathers."[47] In support of his claim for Theophilus's writing being an exemplar for others, Lavallee footnotes a reference to Eusebius's *Ecclesiastical History* 4.24. The actual passage, which Lavallee does not cite, states the following:

> Of Theophilus, whom we have mentioned as bishop of the church of Antioch, three elementary works addressed to Autolycus are extant; also another writing entitled Against the Heresy of Hermogenes, in which he makes use of testimonies from the Apocalypse of John, and finally certain other catechetical books. And as the heretics, no less then than [sic] at other times, were like tares, destroying the pure harvest of apostolic teaching, the pastors of the churches everywhere hastened to restrain them as wild beasts from the fold of Christ, at one time by admonitions and exhortations to the brethren, at another time by contending more openly against them in oral discussions and refutations, and again by correcting their opinions with most accurate proofs in written works. And that Theophilus also, with the others, contended against them, is manifest from a certain discourse of no common merit written by him against Marcion. This work too, with the others of which we have spoken, has been preserved to the present day. Maximinus, the seventh from the apostles, succeeded him as bishop of the church of Antioch.[48]

Nowhere in this passage from Eusebius is there support for anything Lavallee claims about Theophilus's work. There is nothing about its apparent extensive treatment of creation, nor about its paradigm-setting legacy. Yes, Lavallee has properly identified passages in *To Autolycus* that show Theophilus as an early adherent of *creatio ex nihilo*.[49] But in his desire to show that the early church defended creation science, he has read into Eusebius's rather benign statements about Theophilus to support his claim that all the Fathers used him as a model and would, therefore, agree with all that he wrote in *To Autolycus*. His claim is not only inaccurate but also unsupported by his reference.

---

[47]Lavallee, "Early Church Defended Creation Science," 2.
[48]Eusebius of Caesarea, *Church History* 4.24 (NPNF[2] 1:202).
[49]Lavallee cites *To Autolycus* 2.4.

But Lavallee is not finished with his generalizations. He continues by asserting that the "classical scholar and Bible translator" Jerome (347–420) saw fit to include Theophilus in his book *Lives of Illustrious Men*.[50] Lavallee cites a statement from the preface that summarizes Jerome's purpose for listing those he does in the treatise, "those who have published any memorable writing on the Holy Scripture."[51] Immediately following this, Lavallee cites Jerome's later description of Theophilus's writings as "short and elegant treatises well fitted for the edification of the Church."[52] Again, even though Jerome says absolutely nothing about Theophilus's position about creation, nor about his legacy, Lavallee leverages these general claims about Theophilus in order to claim him as support. A closer look at Jerome's *Lives of Illustrious Men* will reveal some difficulties with these generalizations.

In addition to telling us that the treatise is about "those who have published any memorable writing on the Holy Scripture," Jerome also has another overarching purpose. In the preface he explains how he has patterned his *Lives* on Cicero's *Brutus*, in which Cicero (106 BCE–43 BCE) catalogues the Latin orators up to his time. For his part, Jerome wished to do the same for ecclesiastical writers. He then reveals his greater purpose for the writing:

> Let Celsus, Porphyry, and Julian learn, rabid as they are against Christ, let their followers, they who think the church has had no philosophers or orators or men of learning, learn how many and what sort of men founded, built and adorned it, and cease to accuse our faith of such rustic simplicity, and recognize rather their own ignorance.[53]

The impression that Lavallee appears to project onto Jerome's *Lives* as it relates to Theophilus's creation account is that it was both mentioned explicitly and recommended as edifying on that basis. But

---

[50]Jerome, *Lives of Illustrious Men* (NPNF² 3:359-84).

[51]Lavallee, "Early Church Defended Creation Science," 2, citing Jerome, preface to *Lives* (NPNF² 3:359).

[52]Ibid., citing Jerome, *Lives* 25 (NPNF² 3:369).

[53]Jerome, preface to *Lives* (NPNF² 3:359).

neither of these are shown in Jerome's *Lives*. Rather, Jerome recommends all of Theophilus's writings, not only *To Autolycus*, because they have a style and language that is elegant and expressive. This appears to correspond with Jerome's stated purpose in the preface to show that there is no basis to accuse Christianity of "rustic simplicity." Theophilus's works are recommended neither because of his doctrine of *creatio ex nihilo* nor because of the legacy *To Autolycus* has as an exemplar for later Fathers. *All* of Theophilus's writings are recommended because they are evidence that Christianity possesses men of learning.

One Christian included in Jerome's *Lives* is Justin Martyr, who died approximately twenty years before Theophilus. Thus, because Jerome includes Justin in this treatise it is apparent that he published "memorable writing on the Holy Scripture." Of Justin specifically, Jerome states that he "laboured strenuously in behalf of the religion of Christ" and was even martyred for Christ.[54] This, indeed, is high pedigree. But this did not keep Justin from advocating a creation by God from formless matter.[55] So, since Jerome advocates both Theophilus and Justin in the same volume of illustrious men, we cannot conclude that they are included because of their opposite stance on *creatio ex nihilo*.

In fact, Lavallee explicitly misrepresents Justin's teaching on creation out of nothing, using him as one who has *not* been taken "captive through hollow and deceptive philosophy, which depends on human tradition."[56] He identifies Justin as a "philosopher" who turned to Christianity and "found this philosophy alone to be safe and profitable."[57] Lavallee does properly cite Justin here, but his claim immediately following that "the early fathers, like Justin . . . believed the Bible and that God created all things out of nothing" reveals his ignorance of the Father he is claiming for support. In his *First Apology* Justin himself states, "And we have been taught that He in the beginning did of His goodness, for

---

[54]Ibid., 23 (NPNF² 3:368).
[55]Justin Martyr, *The First Apology* 10 (ANF 1:165, 59, 179).
[56]Lavallee, "Early Church Defended Creation Science," 1. Lavallee is here citing Col 2:8.
[57]Ibid., 1. Lavallee is here citing Justin Martyr's *Dialogue with Trypho* 8 (ANF 1:198).

man's sake, create all things out of unformed matter."[58] Later, in the same apology, he argues that Plato actually borrowed from Moses "his statement that God, having altered matter which was shapeless, made the world."[59] Justin then quotes Genesis 1:1-3 to demonstrate the agreement between Moses and Plato regarding creation of the world. Not only has Lavallee completely misrepresented Justin's stance on creation out of nothing, he has also misrepresented his position vis-à-vis philosophy after his conversion. Justin's commitment to Middle Platonism continued even after his conversion to Christianity, albeit in ways that sought to remain true to God's revelation through the Logos, Jesus.[60]

If eisegesis involves reading one's own conclusions into a text, then proof-texting is a tool that is often used to support those conclusions. It may involve bringing one's inappropriate modern presuppositions or conclusions to the text and trying to find the same within the text. The above-mentioned article "Genesis Means What It Says" seeks to do this by "arguing" that Basil of Caesarea advocated a literal reading of Genesis. But the article is not an argument at all. Following an assertion of Basil as proof for the early church's literal interpretation of Genesis is a very short biography of his importance in the history of Christian theology. From there the author arranges proof texts from Basil's homilies on the six days of creation under various creation science–inspired headings. Thus, proof texts are offered to substantiate what Basil believed: creation happened instantaneously and recently; the days of Genesis were twenty-four-hour days; the order of events in Genesis 1 is how creation happened; evolutionary ideas are contrary to Scripture; creation was originally "very good" and completely lacking in evil; and the text is to be understood by its plain meaning, not allegorized. There is no attempt to explain, interact with, or understand any of these passages in the contexts that Basil uses them. In a move that is intended

---

[58]Justin Martyr, *First Apology* 10 (ANF 1:165).

[59]Ibid. 59 (ANF 1:179).

[60]See Craig D. Allert, *Revelation, Truth, Canon and Interpretation: Studies in Justin Martyr's Dialogue with Trypho*, Supplements to Vigiliae Christianae 64 (Leiden: Brill, 2002), 63-121.

to be the icing on the creationist cake, the final heading indicates that Basil believed that to interpret Scripture otherwise is to put ourselves above God, the Holy Spirit, who inspired its writing.

Another example of proof-texting is found in James Mook's above-mentioned article. Victorinus of Pettau (d. 304) is identified as a "literalist" who "affirmed that the first day of creation was divided into 12 hours for day and 12 hours for night."[61] In support of this he cites a passage from Victorinus's fragmentary work, *On the Creation of the World*:

> Even such is the rapidity of that creation; as is contained in the book of Moses, which he wrote about its creation, and which is called Genesis. God produced that entire mass for the adornment of His majesty in six days; on the seventh to which He consecrated it. . . . In the beginning God made the light, and divided it in the exact measure of twelve hours by day and by night. . . . The day, as I have above related, is divided into two parts by the number twelve – by the twelve hours of day and night.[62]

Here Mook has cited parts of the first and second paragraph of the fragment. But one may justifiably wonder whether Mook read the rest of the fragment.

He presents Victorinus as a literalist in the same sense as creation scientists, giving the impression that Victorinus's work is in the same vein. But one does not have to read very far to see that Mook has proof-texted Victorinus to the point of misrepresenting him. Immediately following these first two paragraphs, which Mook has conflated, Victorinus turns to the fourth day of creation (Gen 1:16-17). Victorinus shows no concern for its literalness in his explanation for why the fourth day is called "the Tetras":

> Therefore this world of ours is composed of four elements – fire, water, heaven, earth. These four elements, therefore, form the quaternion of times or seasons. The sun, also, and the moon constitute throughout the space of the year four seasons – of spring, summer, autumn, winter; and

---

[61]Mook, "Church Fathers on Genesis," 29.
[62]As cited by Mook (from ANF 7:341).

these seasons make a quaternion. And to proceed further still from that principle, lo, there are four living creatures before God's throne, four Gospels, four rivers flowing in paradise; four generations of people from Adam to Noah, from Noah to Abraham, from Abraham to Moses, from Moses to Christ the Lord, the Son of God; and four living creatures, *viz.*, a man, a calf, a lion, an eagle; and four rivers, the Pison, the Gihon, the Tigris, and the Euphrates. The man Christ Jesus, the originator of these things whereof we have above spoken, was taken prisoner by wicked hands, by a quaternion *of soldiers*. Therefore on account of His captivity by a quaternion, on account of the majesty of His works, – that the seasons also, wholesome to humanity, joyful for the harvests, tranquil for the tempests, may roll on, – therefore we make *the fourth day* a station or a supernumerary fast.[63]

Victorinus continues this way throughout the entire fragment, connecting numbers as symbols that express a reality not explicitly connected in the Scripture. The fragment is thus not a literal interpretation of Genesis, as Mook leads his readers to believe. Victorinus ultimately comes to see the week recapitulated in many biblical events:

That He might re-create that Adam by means of the week, and bring aid to His entire creation, was accomplished by the nativity of His Son Jesus Christ our Lord. Who, then, that is taught in the law of God, who that is filled with the Holy Spirit, does not see in his heart, that on the same day on which the dragon seduced Eve, the angel Gabriel brought the glad tidings to the Virgin Mary; that on the same day the Holy Spirit overflowed the Virgin Mary, on which He made light; that on that day He was incarnate in flesh, in which He made the land and water; that on the same day He was put to the breast, on which He made the stars; that on the same day He was circumcised, on which the land and water brought forth their offspring; that on the same day He was incarnated, on which He formed man out of the ground; that on the same day Christ was born, on which He formed man; that on that day He suffered, on which Adam fell; that on the same day He rose again from the dead, on which He created light? He, moreover, consummates His humanity in the number seven: of His nativity, His infancy, His boyhood, His youth, His young-manhood, His

---

[63]Victorinus, *On the Creation of the World* (ANF 7:341). Italics in original.

mature age, His death. I have also set forth His humanity to the Jews in these manners: since He is hungry, is thirsty; since He gave food and drink; since He walks, and retired; since He slept upon a pillow; since, moreover, He walks upon the stormy seas with His feet, He commands the winds, He cures the sick and restores the lame, He raises the blind by His speech, – see ye that He declares Himself to them to be the Lord.

The proof-texting employed by Mook is significant because of what he leaves out. Victorinus's purpose is not to exegete the days of Genesis in a literal manner, as Mook wants his readers to believe – but this is how Mook has used him. John Millam, a critic of Mook's use of Victorinus, gives this judgment: "Mook's use of Victorinus to support a calendar-day view shows deficient scholarship and selective quoting. Clearly, Victorinus is far from being a literalist (according to how we use that term today). So he actually does more to undercut Mook's 24-hour day interpretation than he does to support it."[64]

LACK OF KNOWLEDGE OF THE FATHERS

Another significant problem with those who seek to plunder the church fathers for ammunition in the creation/evolution debate is a lack of knowledge of the Fathers, manifested by failing either to read them or to understand the context of their writings and situation.

In an article asking rhetorically if God really could have created everything in six days, AiG president Ken Ham encourages his audience to read Genesis 1 by clearing away all outside influences. "Just let the words of the passage speak to you," he advises.[65] This way, any confusion can be avoided and we can simply read the text. Reading in this manner will allow one to be "really honest" with the text and allow one to agree with the way he reads it. Ham bemoans his impression that

---

[64]John Millam, "Coming to Grips with the Early Church Fathers' Perspectives on Genesis," *Today's New Reason to Believe* (blog), September 22, 2011, https://tnrtb.wordpress.com/2011/09/22/coming-to-grips-with-the-early-church-fathers%E2%80%99-perspectives-on-genesis-part-3-of-5.
[65]Ken Ham, "Could God Really Have Created Everything in Six Days?," September 27, 2007, https://answersingenesis.org/days-of-creation/could-god-really-have-created-everything-in-six-days/.

the majority of Christians in the West do not read it in this way. These people have allowed outside influences to color their pure reading of the text. "In *every* instance," Ham explains, "where someone has *not* accepted the 'days' of creation to be ordinary days, they have not allowed the words of Scripture to speak to them in context, as the language requires for communication. They have been influenced by ideas from *outside* of Scripture."[66]

When he gets to his short section on the church fathers, Ham claims that most of them actually accepted the days of creation as twenty-four-hour days.[67] He recognizes that some Fathers did not teach the days as ordinary days, but – in keeping with his earlier exhortation to read the Bible with no outside influences – the problem is that these Fathers had been influenced by "Greek philosophy," which caused them to interpret the days as allegorical. Apparently Ham, and those who agree with him, *read* the text in an unmediated and uninfluenced way, while all others *interpret* – that is, are influenced by outside sources. For Ham, our own context does not matter as we read, because we can easily escape it. Only the context of the passage comes into play in interpretation.

There are two issues here worth considering. The first is the assumption that reading can be done from a neutral stance or that it is possible (and better) without interpretation. Second, Ham assumes that only the church fathers with whom he agrees managed to escape outside influences. The two issues are, of course, related since the accusation that certain Fathers were wrong in their readings was because they were influenced by outside sources. In other words, they did not just read.

The basis of Ham's assumption is the presupposition that somehow interpretation is wrong. We should simply state what the Bible clearly says rather than allow outside influences to cloud matters with interpretation. Calvin College professor James K. A. Smith, who deals with

---

[66]Ibid. Italics in original.

[67]Ham gives no primary source references here but does cite (with no page numbers), M. Van Bebber and P. Taylor, *Creation and Time: A Report on the Progressive Creationist Book by Hugh Ross* (Mesa, AZ: Films for Christ, 1994).

this issue extensively, relates an example of this kind of attitude that illustrates its ongoing influence in conservative Christianity.[68] Smith explains his discovery of an advertisement for a Bible in a leading evangelical periodical. The dust cover proclaimed "God's Word. Today's Bible translation that says what it means." Underneath the photograph of the Bible, in large bold letters, the publishers added "NO INTERPRETATION NEEDED." Granted, this is a somewhat extreme (albeit true!) example of the misunderstanding that we can somehow surmount our humanity and locatedness and attain a pure reading that delivers "the explicit teaching of Scripture" in an unmediated way. But is this kind of unmediated immediacy to the Bible really possible in our historical particularity? Can we really read the Bible free of any influence or without acknowledging those who have gone before us?

Consider what might be involved in taking Ham's exhortation to just read the text and let it speak to you. The Bibles most of us use are translations from the original Hebrew and Greek. Translation is not simply a matter of finding equivalent words in two languages.[69] The task of transposing material from one world of thought and language to another can be very complicated. These different worlds of thought require the translator to understand both cultures. This means that interpretation is already involved in the task of translation — grammatical and lexical decisions are made that allow the readers of translations to understand the words of the text.

---

[68]James K. A. Smith, *The Fall of Interpretation: Philosophical Foundations for a Creational Hermeneutic*, 2nd ed. (Downers Grove, IL: InterVarsity Press, 2000), 38. For an excellent introduction to hermeneutics, see Jens Zimmermann, *Hermeneutics: A Very Short Introduction* (Oxford: Oxford University Press, 2015).

[69]A good explanation of translation issues as they relate to Genesis is John H. Walton, *The Lost World of Genesis One: Ancient Cosmology and the Origins Debate* (Downers Grove, IL: IVP Academic, 2009), 7-13, 37-38. For a more detailed discussion on the issues translation raises for an evangelical doctrine of Scripture, see Craig D. Allert, "Is a Translation Inspired? The Problems of Verbal Inspiration for Translation and a Proposed Solution," in *Translating the Bible: Problems and Prospects*, ed. Stanley E. Porter and Richard H. Hess, *Journal for the Study of the New Testament Supplement Series* 173 (Sheffield, UK: Sheffield Academic Press, 1999), 85-113.

Even the opening words of the Bible reflect the interpretive decisions that must be made in order to translate. In his book *Genesis 1–11: A Handbook on the Hebrew Text*, Hope College professor Barry Bandstra explains some difficulties in translating the opening words of Genesis.[70] He explains that it is not clear whether the first clause in Genesis 1:1 is a hypotactic expansion or a paratactic expansion.[71] The difference in the choices has implications for whether the first verse should be understood in a temporal sense or not. This leaves the translator with three choices. The NRSV translates Genesis 1:1 as, "In the beginning when God created the heavens and the earth." But in a footnote the translators indicate that there are two other viable options: "when God began to create the heavens and the earth," or "In the beginning God created the heavens and the earth." Similarly, in his commentary on Genesis, Gordon Wenham states that the "stark simplicity of this, the traditional translation ["In the beginning God created"], disguises a complex and protracted debate about the correct interpretation of vv 1-3."[72] He relates four possible understandings of the syntax of the verse:

1. V 1 is a temporal clause subordinate to the main clause in v 2: "In the beginning when God created . . . , the earth was without form. . . ."

2. V 1 is a temporal clause subordinate to the main clause in v 3 (v 2 is a parenthetic comment). "In the beginning when God created . . . (now the earth was formless) God said. . . ."

3. V 1 is a main clause, summarizing all the events described in vv 2-31. It is a title to the chapter as a whole, and could be rendered "In the beginning God was the creator of heaven and earth." What being creator of heaven and earth means is then explained in more detail in vv 2-31.

---

[70]Barry L. Bandstra, *Genesis 1–11: A Handbook on the Hebrew Text*, Baylor Handbooks on the Hebrew Bible (Waco, TX: Baylor University Press, 2008), 41-42.

[71]"Hypotaxis, hypotactic relationship – A pair of clauses in which one clause is dependent upon the other" (ibid., 617). "Parataxis, paratactic relationship – A relationship between two clauses characterized by equality rather than dependency" (ibid., 618).

[72]Gordon J. Wenham, *Genesis 1–15*, Word Biblical Commentary 1 (Nashville: Thomas Nelson, 1987), 11.

4. V 1 is a main clause describing the first act of creation. Vv 2 and 3 describe subsequent phases in God's creative activity.[73]

The first three translations, as Wenham points out, presuppose the existence of matter before the work of creation begins. These options have all had their advocates in the history of interpretation. My point here is not to debate the merits of each clause but rather to indicate that interpretive decisions are made even in the act of translation. And in some cases these interpretive decisions have implications on some important issues. Ken Ham's exhortation to just read the text gets a bit complicated when understood in the context of translation. Most of us read and study translations of the Bible. How does Ham's exhortation to just read the text function in light of these important translational dilemmas?

An obvious response to avoiding this problem would be to go to the original languages.[74] Besides the obvious commitment required for this endeavor, one would have to learn these languages from people who have themselves been trained in a certain tradition of translation theory, which tends toward certain ways of translating word and grammatical constructions. Even those who do not know the original languages and make good use of lexical aids need to understand that these aids are themselves "deposits of accumulated knowledge of earlier scholars."[75]

The point here is that the simple act of reading the Bible, whether in the original language or in a translation, is already a mediated event. This mediation becomes layered when we factor in our own cultural location. The assumption that we can read the Bible, or anything for that matter, from a neutral stance is naive and misguided. Consider the recent experience my wife and I had while watching the movie *The Secret Life of Pets*.[76]

---

[73]Ibid.

[74]The following is adapted from Stephen R. Holmes, *Listening to the Past: The Place of Tradition in Theology* (Grand Rapids: Baker, 2002), 6-7.

[75]Ibid., 6.

[76]Brian Lynch, Cinco Paul, and Ken Daurio, *The Secret Life of Pets*, directed by Chris Renaud and Yarrow Cheney (Universal City, CA: Universal Pictures, 2016).

Near the beginning of the animated film several vignettes show common reactions of pets whose owners have left to attend to matters outside the home. The pets are portrayed as utterly perplexed about why the owners would leave them alone. It may sound silly, but these vignettes affected my wife and me in a way we did not expect. In our own home we have a nine-year-old Maltipoo dog named Josie, and we could not help but watch this movie through the eyes of our relationship with her. In fact, we went away on vacation several days after viewing the movie. During our drive through the Fraser Canyon in beautiful British Columbia, my wife said something like, "I can't stop thinking about Josie because of that movie. I feel bad we left her." Now we are not so naive as to think that what was portrayed on the screen *actually* happens when we leave Josie, but we were still affected because we know that when we do leave her at home and go out she misses us and does not understand why we leave. We could not watch *The Secret Life of Pets* without thinking about our own dog, and afterward we could not think about our own pet without thinking about the experience of viewing that movie.

Something similar happens when we read the Bible. Whether we realize it or not, we cannot read it without incorporating what has gone on in our own lives, and, hopefully, if we are listening, we should not go about our own lives without thinking about our experience in reading the Bible. Taking Ham's advice to clear our minds and volitionally put aside influences is not only impossible, it (perhaps unwittingly) becomes an attempt to overcome our humanity, which actually ends up devaluing God's creation.[77] This is not to say that all outside influences are good. But we should not try to escape our own historical locatedness – something from which we simply cannot escape.[78]

Ham assumes that we can simply read the passage with no outside influences – that is, unmediated. Smith calls this a claim to immediacy, which assumes, as Ham does, that we can choose whether to be influ-

---

[77]Smith, *Fall of Interpretation*, 38-39, 51.
[78]Holmes, *Listening to the Past*, 6-7.

enced by something. But is this accurate? Does not our commitment to the Bible as God's Word occur within some sort of influential community of faith? Is Ham willing to admit that even his call to just read the Bible comes from within a community of interpretation that protests any understanding of the days of creation as anything other than twenty-four hours? Is he willing to admit that the interpreters he favors come from a community that has a certain stake in interpreting Genesis in a certain way? Does this, perhaps, influence him? Failure to recognize and admit this can actually be a dangerous thing. It can lead to a self-understanding that you actually speak for God, and everyone else had better fall in line. So, when Ham links Romans 3:4 ("Although everyone is a liar, let God be proved true") with his understanding of the days of Genesis, it is hard not to conclude that he thinks he is speaking for God. This can be a scary thing, as John Caputo explains:

> For what we always get — it never fails — in the name of the Unmediated is someone else's highly mediated Absolute: their Jealous Yahweh, their righteous Allah, their infallible church, their absolute Geist that inevitably speaks German. In the name of the Unmediated we are buried in an avalanche of mediations, and sometimes just buried, period. Somehow this absolutely absolute always ends up with a particular attachment to some historical, natural language, a particular nation, a particular religion. To disagree with someone who speaks in the name of God always means disagreeing with God. Be prepared to beat a hasty retreat. The unmediated is never delivered without massive mediation.[79]

If we think we are ignoring all that has gone before us, and coming to the biblical text in an unmediated form, we are deceiving ourselves. We simply cannot do it, nor should we try! For Ham, proper Bible reading requires knowledge of the biblical context. But if it is a given that the Bible was written from a particular context about which we must be aware, surely we must be aware of our own context. Ham

---

[79]John D. Caputo, "How to Avoid Speaking of God: The Violence of Natural Theology," in *Prospects for Natural Theology*, ed. Eugene Thomas Long (Washington, DC: University of America Press, 1992), 129-30, cited in Smith, *Fall of Interpretation*, 48.

seems to think some can escape that context whereas others cannot. Those who cannot are, at best, guilty of poor reading. Is this really possible? Don't we actually *need* help in reading the Bible?

At this point someone may ask about the principle of the clarity (perspicuity) of Scripture. Surely, the argument goes, this is a bedrock principle of conservative Protestantism that supports Ham's assertion. But the Protestant principle of the clarity of Scripture was not a claim that we can and should simply read the Bible and clear our minds of all influences.[80] Even this principle cannot be lifted from its own historical context. Protestants from Luther to Wesley found the perspicuity of Scripture an effective banner to unfurl when attacking Catholics, but it was always a bit troublesome when laypeople began taking the teaching seriously.[81] For the Reformers, popular translations of the Bible did not imply that people were to understand the Scriptures apart from ministerial guidance. So, when dealing with the Catholic scholar Erasmus, Luther could champion the perspicuity of Scripture by stating: "Who will maintain that the public fountain does not stand in the light, because some people in the back alley cannot see it, when every boy in the market place sees it quite plainly?" But when he was confronted with Protestants he called sectarians, Luther admitted the danger of proving anything from Scripture: "I learn now that it is enough to throw many passages together helter skelter, whether they are fit or not. If this be the way, then I can easily prove from the scriptures that beer is better than wine."[82] John Calvin had a similar understanding: "I acknowledge that Scripture is a most rich and inexhaustible fountain of all wisdom, but I deny that its fertility consists in various meanings which any man, at his pleasure may assign."[83]

---

[80]What follows draws on Craig D. Allert, *A High View of Scripture? The Authority of the Bible and the Formation of the New Testament Canon*, Evangelical *Ressourcement* Series (Grand Rapids: Baker Academic, 2007), 174-75.

[81]Nathan O. Hatch, "*Sola Scriptura* and *Novus Ordo Seculorum*," in *The Bible in America: Essays in Cultural History*, ed. N. O. Hatch and M. A. Noll (New York: Oxford University Press, 1982), 61.

[82]Both quotations are cited in Hatch, "*Sola Scriptura*," 61.

[83]John Calvin, *Commentary on Galatians*, in *Calvin's Commentaries: The Epistles of Paul the Apostle to*

Protestant Reformers like Luther and Calvin were affirming something that was recognized first by the church fathers: Scripture could be interpreted in different ways. The Bible is and should be the ultimate authority for Christians, but just because this authority exists does not mean that it will be interpreted in the same way by all Christians.[84] The history of Christian doctrine is not simply a repetition of scriptural statements, as if they need no interpretation. Individuals branded as heretics by the Fathers of the early church invariably took their stand on Scripture, often claiming it as the sole court of appeal. But their interpretations were ruled out of bounds. So when both "sides" appeal to Scripture, what determines a proper and improper reading? Will the exhortation to "just let the words of the passage speak to you" really help?[85]

Athanasius's fourth-century battle with Arian interpretations of Scripture is illustrative of the issues here.[86] There is no doubt that Athanasius held Scripture as the ultimate authority for the Christian and, therefore, all theological truth. But the Arians also agreed with this. So the battle with the Arians was "not a battle *for* the Bible, but a battle *over* the Bible."[87] The Arians, as Athanasius claims throughout his *Four Discourses Against the Arians*, use scriptural terms but fail to accept the truth of Scripture and merely "array" themselves in scriptural language.[88] For Athanasius, one can use Scripture but still miss its meaning. He would likely agree with the medieval saying in reference to the Bible: "This is the book in which everyone looks for his own convictions, and likewise everyone finds his own convictions."[89]

---

the Galatians, Ephesians, Philippians, and Colossians, ed. D. W. Torrance and T. F. Torrance (Grand Rapids: Eerdmans, 1974), 84-85.

[84]This is possibly one of the reasons why a closed New Testament canon was not explicitly deliberated until well into the fourth century. See Allert, *A High View of Scripture?*

[85]Ham, "Could God Really?"

[86]What follows draws on Peter J. Leithart, *Athanasius* (Grand Rapids: Baker Academic, 2011), 33-39.

[87]Ibid., 33. Italics added.

[88]*Four Discourses Against the Arians* is found in NPNF² 4:303-447. See also Leithart, *Athanasius*, 34.

[89]Cited in Yves M.-J. Congar, *Tradition and Traditions: The Biblical, Historical, and Theological Evidence for Catholic Teaching on Tradition* (San Diego: Basilica, 1966), 385.

Of course, one can see that Athanasius's argument cuts both ways. What makes his understanding of Scripture proper and the Arians' improper? Was it because Athanasius "let the words of the passage speak" to him? What counts as a legitimate reading of Scripture? In the context of the Arian crisis, Athanasius appeals to the heritage of the Christian church – to the Fathers from whom he says the Arians have deviated.[90] The Arians were guilty of deviating from what had been faithfully "transmitted from father to father."[91] They accorded a meaning of Scripture that does not reflect the mind of the faithful, but a "misinterpretation, according to their private sense."[92]

Arians took passages like Proverbs 8:22 as indication that Jesus was a created being.[93] Of course, Athanasius disagreed that the passage can be used to show this, stating that it is necessary to "unfold the sense of what is said, and to seek it as something hidden, and not nakedly to expound as if the meaning were spoken 'plainly,' lest by a false interpretation we wander from the truth."[94] Athanasius was aware of influences and actually pointed Christians to the importance of using the proper ones. He realized his interpretation of Scripture stood in a long heritage of influence, and he appealed to it as proper and orthodox.

Ham names the influence of "Greek philosophy" as the culprit causing some church fathers to interpret the days of Genesis as other than twenty-four-hour days.[95] The naiveté of this statement is apparent to those with even a mere acquaintance with the church fathers. Does this mean that those Fathers who interpreted the days of Genesis as twenty-four hours were not influenced by philosophy?

---

[90]See Athanasius, *Defence of the Nicene Definition* 6.25-27 (NPNF[2] 4:166-69).

[91]Ibid. 6.27 (NPNF[2] 4:168).

[92]Athanasius, *Four Discourses Against the Arians* 1.11.37 (NPNF[2] 4:327).

[93]As cited by Athanasius in *Discourse* 2.19.44 (NPNF[2] 4:372): "The Lord created me a beginning of His ways, for His works."

[94]Ibid.

[95]Simon Turpin makes the same accusation in "Review of Lennox's Book *Seven Days That Divide the World: The Beginning According to Genesis and Science*," June 27, 2012, https://answersingenesis.org/reviews/books/review-of-seven-days-that-divide-the-world-john-lennox/.

Ham's assertion about the influence of philosophy could actually be used against him. The approach to the biblical text advocated by Ham above is heavily indebted to Enlightenment philosophers like René Descartes (1596–1650) and John Locke (1632–1704), who claimed that we could actually surmount our own context and attain pure readings of texts and reality.[96] In fact, evangelical historian Mark Noll argues that virtually every aspect of the evangelical attachment to the Bible was shaped by the Enlightenment.[97] Thus, even though Ham may not know it or admit it, he is also influenced by some form of philosophy, among other things.

Since Ham's article does not devote much space to the issue of influences, some may complain that my relatively long critique of it here is overblown or even unfair. After all, Ham is no expert in theology, history, or philosophy. But this is just my point. He has certain foundational assumptions that subsequently influence his conclusions. Admitting these assumptions could allow for a better way forward. I am not interested in fighting a battle here. I am interested in offering an approach to the Fathers that respects and understands the context within which they worked and that draws on scholars who have devoted years to understanding and explaining this complex context rather than the parachute approach that simply drops in and selectively rummages for data in support of one position.

---

[96]Smith, *Fall of Interpretation*, 39. I will forgo the temptation to delve further into this issue since it is too far afield of my intent. For more on this important issue, see D. W. Bebbington, "Evangelical Christianity and the Enlightenment," *Crux* 25 (1989): 29-36; D. W. Bebbington, "Evangelical Christianity and Modernism," *Crux* 26 (1990): 2-9; Philip D. Kenneson, "There's No Such Thing as Objective Truth, and It's a Good Thing, Too," in *Christian Apologetics in the Postmodern World*, ed. Timothy R. Phillips and Dennis L. Okholm (Downers Grove, IL: InterVarsity Press, 1995), 155-70; George M. Marsden, *Fundamentalism and American Culture: The Shaping of Twentieth Century Evangelicalism, 1870–1925* (Oxford: Oxford University Press, 1980); Mark A. Noll, *Between Faith and Criticism: Evangelicals, Scholarship, and the Bible in America*, 2nd ed. (Grand Rapids: Baker, 1991); Mark A. Noll, "Evangelicals, Creation, and Scripture: Legacies from a Long History," *Perspectives on Science and Christian Faith* 63, no. 3 (2011): 147-58; and Timothy P. Weber, "The Two-Edged Sword: The Fundamentalist Use of the Bible," in *The Bible in America: Essays in Cultural History*, ed. N. O. Hatch and Mark A. Noll (Oxford: Oxford University Press, 1980), 101-20.

[97]Noll, *Between Faith and Criticism*, 97.

James Mook is somewhat more sophisticated in his discussion of philosophic influences on the Fathers, but he still begins with unwarranted assumptions and ends with incorrect conclusions. In describing the "naturalistic milieu of the Fathers," Mook uses several early church leaders to instruct us on how we should think about philosophy's influence on Christianity and, specifically, cosmology.

He cites Hippolytus (ca. 170–ca. 235), Basil of Caesarea (329–379), and Lactantius (250–325) as proof that "the fathers asserted their views in large part to refute Greek philosophy's naturalistic theories of origins, which were similar to modern ideas."[98] Mook especially uses Basil, asserting that his "words about the temporary life of naturalistic theories should be considered when we use current scientific theories of origins as epistemic foundations for interpreting Scripture."[99] The problem here is the sweeping generalizations Mook makes based on his proof-texting of the three Fathers. The claim that the Fathers refuted certain naturalistic theories of origins is not at issue here. The issue is the Fathers Mook uses to show this and the question of whether these examples warrant the full-scale rejection of science and philosophy as he appears to infer.

According to Mook, Basil's frequent allusions to the philosophers and their cosmologies is best exemplified by the words in his first homily on the Hexaemeron:

> "In the beginning God created the heaven and the earth." I stop struck with admiration at this thought. . . . The philosophers of Greece have made much ado to explain nature, and not one of their systems has remained firm and unshaken, each being overturned by its successor. It is vain to refute them; they are sufficient in themselves to destroy one another. Those who were too ignorant to rise to a knowledge of a God, could not allow that an intelligent cause presided at the birth of the Universe; a primary error that involved them in sad consequences. Some had recourse to material principles and attributed the origin of the Universe to

---

[98]Mook, "Church Fathers and Genesis," 28.
[99]Ibid., 28n11.

the elements of the world. Others imagined that atoms, and indivisible bodies, molecules and ducts, form, by their union, the nature of the visible world. Atoms reuniting or separating, produce births and deaths and the most durable bodies only owe their consistency to the strength of their mutual adhesion: a true spider's web woven by these writers who give to heaven, to earth, and to sea so weak an origin and so little consistency! It is because they knew not how to say "In the beginning God created the heaven and the earth." Deceived by their inherent atheism it appeared to them that nothing governed or ruled the universe, and that all was given up to chance. To guard us against this error the writer on the creation, from the very first words, enlightens our understanding with the name of God; "In the beginning God created."[100]

Basil claims here that each philosophical theory is overturned by succeeding views and that everything is attributed to "chance." This is enough for Mook to conclude that Basil opposed any Greek philosophy on the matter of creation and that this should serve as a warning for us when we use current scientific theories of origins as epistemic foundations for interpreting Scripture.

Once again, Mook has lifted one part of an extended argument from Basil with little regard for its context or purpose. I deal with the extended section, which is essentially the entire first homily Basil preached on the Hexaemeron, in detail in chapter six. Basil's concern up until section seven of this first homily has been to show that creation is not coeternal with God, and that God created the invisible world before the "beginning of Genesis 1." In section eight he explains that inquiring into creation, whether it is visible or invisible, would require too much time. The necessity of such an inquiry is obviated because "a concern for these things is not at all useful for the Church."[101] This claim would have reminded his listeners of something he stated at the beginning of the homily – a section not cited by Mook. In Homily 1.1, Basil claims that the

---

[100]Basil, *Hexaemeron* 1.2, as cited by Mook, "Church Fathers and Genesis," 27-28. Mook uses the NPNF[2] vol. 8 translation.

[101]Basil, *Hexaemeron* 1.8 (FC 46:14).

narrative of the Hexaemeron was expressed in the "teachings of the Spirit," for "the salvation of those taught." This is what Basil sees the first chapter of Genesis addressing, and it is where he directs his homilies. This is why he is content to let the words of Isaiah and Psalms[102] stand as they are and not press them to say something scientific. He states, "Therefore, I urge you to abandon these questions and not to inquire upon what foundation it stands. If you do that, the mind will become dizzy, with the reasoning going up to no definite end."[103]

From section eight of Basil's first homily on the Hexaemeron and through to the end (section eleven), he refers to and even employs arguments and conclusions from the Greek philosophy he supposedly rejects.[104] Here he offers explanations that "some of the inquirers into nature" give about the nature and the form of the earth. It is striking, however, that rather than refuting these arguments, which is what we would expect if Mook's reading is accurate, Basil indicates their plausibility and encourages his hearers to praise God for ordering the world in such a way:

> And should any of these things which have been said seem to you to be plausible, transfer your admiration to the wisdom of God which has ordered them so. In fact, our amazement at the greatest phenomena is not lessened because we have discovered the manner in which a certain one of the marvels occurred. But, if this is not so, still let the simplicity of faith be stronger than the deductions of reason.[105]

In the final section (11) of the homily Basil discusses more theories from these inquirers, this time concerning the heavens. His point when

---

[102]Basil, *Hexaemeron* 1.8 (FC 46:14), Is 51:6 (LXX): "He established the heaven as if smoke"; Is 40:22 (LXX): "He that stretcheth out the heavens as a vaulted ceiling." Ibid., 1.9 (FC 46:15-16), Ps 74:4 (LXX): "I have established the pillars thereof"; Ps 23:2 (LXX): "He hath founded it upon the seas."

[103]Basil, *Hexaemeron* 1.8 (FC 46). Cf. Basil, *Hexaemeron* 1.9 (FC 46:15), "Set a limit, then, to your thoughts, lest the words of Job should ever censure your curiosity as you scrutinize things incomprehensible, and you also should be asked by him: 'Upon what are its bases grounded?' [Job 38:6]."

[104]He employs Aristotle, *On the Heavens* 2.13.294b, 294a, 2.13.295b, 296a; and Plato, *Timaeus* 31b.

[105]Basil, *Hexaemeron* 1.10 (FC 46:17).

this short discussion is completed is that it is all "idle chatter."[106] The philosophers refute one another when speaking about the substance of the earth, so Basil exhorts his hearers to take refuge in what Moses so clearly affirms, that "God created the heavens and the earth."[107] The issue for Basil is not that all Greek philosophy is bad. If this were true, he would not have admitted the plausibility of some of it and encouraged his hearers to "transfer your admiration to the wisdom of God which has ordered them so." The issue is that creation points us to God.[108] We have already seen Basil's commitment in his homilies on Genesis 1 toward the salvation of his hearers.[109] A short time later he builds on this commitment by explaining that the world is a "place of training and a school for all the souls of men."[110] He concludes the entire homily in the same vein:

> Let us glorify the Master Craftsman for all that has been done wisely and skillfully; and from the beauty of the visible things let us form an idea of him who is more than beautiful; and from the greatness of these perceptible and circumscribed bodies let us conceive of him who is infinite and immense and who surpasses all understanding in the plentitude of his power. For, even if we are ignorant of things made, yet at least, that which in general comes under our observation is so wonderful that even the most acute mind is shown to be at a loss as regards the least of the things in the world, either in the ability to explain it worthily or to render due praise to the creator, to whom be all glory, honor, and power forever. Amen.[111]

The distinction being made here is subtle but very important if we are to understand why Mook's use of this passage is misleading. It is true that Basil rejects theories of creation that are devoid of God as the

---

[106]Basil, *Hexaemeron* 1.11 (FC 46:19).

[107]Ibid.

[108]For more on this aspect of the church fathers' (including Basil) approach to creation, see Paul M. Blowers, "Entering 'This Sublime and Blessed Amphitheatre': Contemplation of Nature and Interpretation of the Bible in the Patristic Period," in *Nature and Scripture in the Abrahamic Religions,* vol. 1, *Up to 1700,* ed. Scott Mandelbrote and Jitse M. van der Meer, Brill's Series in Church History (Leiden: Brill, 2008), 147-74.

[109]Basil, *Hexaemeron* 1.1 (FC 46:5).

[110]Ibid., 1.5 (FC 46:9).

[111]Ibid., 1.11 (FC 46:19).

Creator. But we must not take this to mean that these theories are necessarily rejected wholesale. I think that Basil's encouragement in 1.10 to accept plausible explanations of origins and turn them to God for his ordering of them in this way is a strong indication of this. But the theories are merely a secondary concern for Basil. Scripture does not delve into the details of creation, and Basil even wonders why we do. Theories are just that – attempts to understand what cannot be understood. Each one is refuted by the other. It is enough for Basil to see creation as a training ground where God's creation turns us back to the Creator himself in praise and awe. Andrew Louth states it well in his comments about Basil's relationship to the science and philosophy of his day, which he knew very well. He discusses astronomical and calendrical matters and expresses the church's opposition to astrology.[112] He knows much about the different kinds and habits of plants, reptiles, fish, birds, and animals that are drawn from science and philosophy. Louth continues:

> The notes in the critical edition [of Basil's *Hexaemeron*] are full of references to classical parallels – to the elder Pliny, for instance, not that Basil could have used Pliny's *Natural History*, but because it is the most extensive to survive from antiquity, and also to Aristotle and, interestingly, to the great medical doctor Galen, both of whose works Basil may well have known and used. When he comes to discuss human beings his picture . . . draws heavily on Plato's *Timaeus*. For Basil, there was no opposition between scripture and science; he used contemporary science to fill the bare outline of the cosmos found in Genesis. He can do this because he is quite clear that the scriptural account is not a scientific account. It is no criticism of Moses, he remarks, that he did not clarify whether the earth is a sphere, a cylinder, a disc, or like a great basket, hollowed out in the middle, or that he did not give the measurement of the circumference of the earth. All these things . . . are irrelevant to Moses' purpose: to proclaim God as Creator, the cosmos as his good creation, and the place of the human in all of this [*Hexaemeron* 9.1].[113]

---

[112]Ibid., 6.4-8 (FC 46:88-97).
[113]Louth, "Six Days of Creation," 48-49.

The crowning argument in Mook's attempt to prove the church fathers were young-earth creationists is his claim that they all taught an earth that was less than six thousand years old. He does this through something he calls the "sex/septa millennial view," which, he states, was held by all the Fathers.[114] In the first two centuries of Christian history, Mook argues, the Fathers held a premillennial eschatology in which the seventh age would be the millennium. A shift occurred in this belief sometime after the second century when the predominant view became amillennial. The shift, however, did not change the espousal of the six-thousand-year schema of history.

Mook explains that the sex/septa millenary view is based on Psalm 90:4 ("For a thousand years in your sight are like yesterday when it is past") and 2 Peter 3:8 ("With the Lord one day is like a thousand years, and a thousand years are like one day"). The Fathers, he asserts, believed that each day of creation was a type for a period of one thousand years in the future history of the church. Mook does little more to describe the context he has entered here but chooses, rather, to go directly to discussing the Fathers who held the view.

What Mook calls the sex/septa millenary view, others have called the "creation-week typology" or the "cosmic-week theory."[115] Its foundation for early Christianity is found in Revelation 20:1-10, which was interpreted by some to indicate an earthly reign of one thousand years after the return of Christ. According to this typology, "history will run a course of six one-thousand year 'days' after which Christ will return to usher in the eschatological Sabbath."[116]

Mook wisely refrains from claiming that this view is anchored only in Revelation 20. It actually had less to do with the book of Revelation

---

[114]Mook, "Church Fathers and Genesis," 38-48.
[115]Regarding "creation-week typology," see David G. Dunbar, "The Delay of the Parousia in Hippolytus," *Vigiliae Christinae* 37, no. 4 (1983): 315. For more on the "cosmic-week theory," see Charles E. Hill, *Regnorum Caelorum: Patterns of Millennial Thought in Early Christianity*, 2nd ed. (Grand Rapids: Eerdmans, 2001), 161.
[116]Dunbar, "Delay of the Parousia," 315.

than we might think. Bernard McGinn explains that this millennialism was "a theological construction that sought to amalgamate many sources, both from the Bible and from other writings and oral traditions, some not readily available today."[117] But unfortunately, Mook's general claim that the church fathers had a premillennial eschatology in the first two centuries is overstated. The fact is, we simply do not know how widespread these millennial, or chiliastic, beliefs were among Christians for the first one and a half centuries of the church's existence.[118] Charles Hill has devoted an entire volume to showing that there were both millennial and nonmillennial eschatological beliefs. Thus, "a solidly entrenched and conservative, non-chiliastic eschatology was present in the Church to rival chiliasm from beginning to end."[119] This means "it is incorrect to claim that chiliasm was an integral feature of all Christian apocalypticism."[120]

Mook's generalization that all the Fathers had a premillennial view not only obscures an important reality about early Christianity but also should cause a bit more deliberation about his broader argument. He paints with a broad stroke regarding the unity of millennial views in the first two centuries of Christianity, which aids in his subsequent assertions. If everyone in the early church held this view, then this gives weight to his later arguments based on the apparent widespread agreement. But what if there was no widespread agreement?

---

[117]Bernard McGinn, "Turning Points in Early Christian Apocalypse Exegesis," in *Apocalyptic Thought in Early Christianity*, ed. Robert J. Daly (Grand Rapids: Baker Academic / Brookline, MA: Holy Cross Orthodox Press, 2009), 84. Identifying and discussing these sources is beyond the scope of this study. For more on this, see David Frankfurter, "The Legacy of Jewish Apocalypses in Early Christianity: Regional Trajectories," in *The Jewish Apocalyptic Heritage in Early Christianity*, ed. James C. VanderKam and William Adler (Assen: Van Gorcum / Minneapolis: Fortress, 1996), 129-200; and Richard Landes, "Lest the Millennium Be Fulfilled: Apocalyptic Expectations and the Pattern of Western Chronography, 100–800 CE," in *The Use and Abuse of Eschatology in the Middle Ages*, ed. Werner Verbeke, Daniel Verhelst, and Andries Welkenhuysen (Leuven: Leuven University Press, 1988), 137-211.

[118]Latin *mille* = *thousand*; Greek *chilias* = thousand. See also McGinn, "Turning Points," 84.

[119]Hill, *Regnorum Caelorum*. Jaroslav Pelikan states that millennialism "stood in tension with other descriptions of the reign of Christ" (*The Christian Tradition: A History of the Development of Doctrine*, vol. 1, *The Emergence of the Catholic Tradition [100–600]* [Chicago: University of Chicago Press, 1971], 124).

[120]McGinn, "Turning Points," 86.

Even one of the sources Mook uses to support his assertion of widespread agreement contradicts his generalization. Mook is correct that Justin Martyr was a millennialist. But his citation of Justin as arguing that "right minded Christians" believe in a thousand-year earthly reign in Jerusalem is not the whole picture.[121] Mook does not say it, but it appears that he is trying to link Justin's claim with the orthodox position, thus giving the further impression of widespread agreement *and* orthodoxy. But earlier in the very same chapter of *Dialogue with Trypho*, Justin makes the opposite claim. Trypho, a Jew, asks Justin if he is, in fact, a millennialist. Justin answers that he "and many others are of this opinion."[122] But Justin also sees fit to add that "many who belong to the pure and pious faith, and are true Christians, think otherwise."[123] So, Justin himself tells us that in the latter half of the second century this was not an issue of orthodoxy. Reading the entirety of chapter eighty shows that the issue for Justin was that some nonchiliasts (not all) also rejected the resurrection of the body, which was essential to orthodoxy. But Justin's position on the millennium was one permissible within a range of others.[124]

We also see some inconsistencies in Mook's presentation of Hippolytus.[125] Mook cites Hippolytus's commentary on Daniel as proof for his agreement with the cosmic-week typology.[126] Assuming for a

---

[121]Mook, "Church Fathers and Genesis," 40, citing Justin Martyr's *Dialogue with Trypho* 80. Mook incorrectly cites his references to Justin. In his two quotations of *Dialogue with Trypho*, Mook refers his readers to ANF 1:80, 81. Apparently Mook has mixed page numbers with chapter numbers, because the passages he cites are from chapters eighty and eighty-one. The correct page numbers in ANF vol. 1 are 240-41.

[122]Justin Martyr, *Dialogue with Trypho* 80 (ANF 1:239).

[123]Ibid.

[124]Pelikan, *Christian Tradition*, 1:125.

[125]Since the debate concerning the identity of Hippolytus is not relevant to the topic at hand, I will refrain from discussing the issue. For more on this, see sources indicated in Hill, *Regnum Caelorum*, 160-61; and McGinn, "Turning Points," 89-91.

[126]Once again, Mook has given an incorrect reference. He lists Hippolytus's fragmentary commentary on Daniel as being in ANF volume three. The correct reference is Hippolytus, *On Daniel* 2.4 (ANF 5:179), as cited by Mook, "Church Fathers and Genesis," 42-43: "For 'a day with the Lord is as a thousand years.' Since, then, in six days God made all things, it follows that 6,000 years must be fulfilled. And they are not yet fulfilled, as John says: 'five are fallen; one is,' that is, the sixth; 'the other is not yet come.'"

moment that Hippolytus does subscribe to a literal cosmic-week pattern (more on this below), Mook's use of him, once again, is suspect in terms of what he leaves out. He cites a section in Hippolytus that he claims anchors him squarely as a young-earth creationist.[127] But it is important to consider the broader context within which the quote used by Mook sits. Hippolytus believes it is important to deal with the "matter of times." He continues:

> For the first appearance of our Lord in the flesh took place in Bethlehem, under Augustus, in the year 5500; and He suffered in the thirty-third year. And 6,000 years must needs be accomplished, in order that the Sabbath may come, the rest, the holy day "on which God rested from all His works." For the Sabbath is the type and emblem of the future kingdom of the saints, when they "shall reign with Christ," when He comes from heaven, as John says in his Apocalypse: for "a day with the Lord is as a thousand years." Since, then, in six days God made all things, it follows that 6,000 years must be fulfilled. And they are not yet fulfilled, as John says: "five are fallen; one is," that is, the sixth; "the other is not yet come."[128]

But immediately following the section cited by Mook, Hippolytus mentions some who might ask for proof of Jesus' birth in the year 5500, which Hippolytus had just specified. To answer the query, Hippolytus uses a nonliteral interpretation of biblical stories, such as in the dimensions of the ark of the covenant.[129] The five and a half cubits of the ark symbolize five and a half millennia, concluding in Jesus' birth. Hippolytus links the ark of the covenant with the incorruptible ark of Christ's body, which was made manifest through his birth.[130] "From the birth of Christ, then," reasons Hippolytus, "we must reckon the 500 years that remain to make up the 6,000, and thus the end shall be." That Christ appeared at the end of 5,500 years with this incorruptible ark of

---

[127]Mook cites Hippolytus, *On Daniel* 2.4 (ANF 5:179).

[128]Hippolytus, *On Daniel* 2.4 (ANF 5:179).

[129]I have used Dunbar's helpful summary of this interpretation from "Delay of the Parousia," 315-16.

[130]Hippolytus, *On Daniel* 2.6 (ANF 5:179).

his body is shown in John 19:14 ("Now it was the sixth hour"),[131] which means one half day. Since a day with the Lord is one thousand years, half of that is five hundred years.

The irony here is that Mook relies on a source that employs a nonliteral reading of certain biblical passages. Yet reliance on this interpretation is employed in order to prove a young-earth reading of Genesis. What Mook leaves out of his citation of Hippolytus should at least lead us to question the appropriateness of citing nonliteral interpretations in support of literal readings. One could justifiably accuse Mook of selectivity in his use of the sources.

Another example from Hippolytus illustrates the difficulty of locating him solidly in the position Mook asserts. Sometime near the beginning of the third century CE a presbyter of the church at Rome named Caius (also known as Gaius) was actively writing. We have only fragments of his work, and most of what we know about him comes from Eusebius and Hippolytus. Hippolytus disagreed with Caius's teaching about the authenticity of the book of Revelation that he presented in his *Dialogue* against the Montanist leader Proclus. Our knowledge about Caius was enhanced near the end of the nineteenth century when John Gwynn discovered a twelfth-century document written by Dionysius Bar Salibi called *Commentary on the Apocalypse, Acts and Epistles*.[132] Bar Salibi's commentary is significant in our context because it contains excerpts from Hippolytus's exegesis of Revelation 20. In 1888 John Gwynn published an article with the Syriac-English translations of the five passages discussing Hippolytus's position with respect to Caius.

The fifth passage is relevant to our topic here because in it Hippolytus touches on the subject of millenarian prediction. Recall that Mook cites Hippolytus's comments from his commentary on Daniel as

---

[131]As cited by Hippolytus.
[132]John Gwynn, "Hippolytus and His 'Heads Against Caius,'" *Hermathena* 6, no. 14 (1988): 397-418.

supporting the cosmic-week typology.[133] But the Father's comments about Revelation 20 cast a doubt on Mook's conclusion:

> And the number of the years is not the number of days, but it represents the space of one day, glorious and perfect. . . . "This is the day which the Lord hath made." . . . Accordingly, when with the eye of the spirit John saw the glory of that day, he likened it to the space of a thousand years; according to the saying, "One day in the world of the righteous is as a thousand years." And by the number he shows that day to be perfect, for those that are faithful.[134]

Commenting on this passage, Charles Hill states:

> For this Hippolytus, one day does not equal a thousand years; rather, a thousand years equal one day. That is, the thousand years of Rev. 20 is a figure, a symbol for a single, glorious, and perfect day, and this day has no appearance of belonging to a world-week chronology. This instead is a complete reversal of the normal cosmic-day idea.[135]

Rather than providing us with support for Hippolytus's literal understanding of the millennium, as Mook asserts of him, Hippolytus's refutation of Caius's teachings "provides evidence for an emerging non-literal view of the predictions found in the Apocalypse."[136]

In light of this, one wonders if Mook would take his own advice to those who misrepresent the Fathers. Recall Mook's four important questions that should be asked of scholars using the Fathers inappropriately in the service of deep time:

> First, which specific ancient treatises were these modern scholars using to class the ancients into such post-Darwinian sounding categories? Second, were there any treatises or resources these modern writers overlooked? Third, if there were overlooked resources, was this innocent oversight due to perhaps consulting only secondary sources? And fourth, if these men were presented with sufficient patrological counter-evidence, would they acknowledge this in subsequent writings?[137]

---

[133]Hippolytus, *On Daniel* 2.6 (ANF 5:179).
[134]Gwynn, "Hippolytus," 403-4.
[135]Hill, *Regnum Caelorum*, 165.
[136]McGinn, "Turning Points," 92.
[137]Mook, "Church Fathers on Genesis," 24.

Did Mook overlook any ancient sources? If so, was it an innocent oversight or an over-reliance on secondary sources? Would this information ("counter-evidence") about Hippolytus cause any change in his presentation?

Preparing and writing this chapter has been another lesson for me on how some quarters of evangelicalism tend to handle disagreements and controversies. What has been seen above in some creation science appropriations of the Fathers is the tendency toward proof-texting with little to no regard for context. The results are a misappropriation of the Fathers. I applaud those who desire to show the relevance of the Fathers to the contemporary church. But when it is done at the expense of their own context and concerns, they are being misused and misappropriated. When this is done, it is difficult not to conclude that ideology is guiding their appropriation. Serious interaction with the Fathers is necessary, but this takes time, deliberation, and patience.

RECOMMENDED READING

Millam, John. "Coming to Grips with the Early Church Fathers' Perspectives on Genesis." *Today's New Reason to Believe* (blog), September 22, 2011. https://tnrtb .wordpress.com/2011/09/22/coming-to-grips-with-the-early-church-fathers%E2% 80%99-perspectives-on-genesis-part-3-of-5.

Williams, D. H. *Retrieving the Tradition & Renewing Evangelicalism: A Primer for Suspicious Protestants*. Grand Rapids: Eerdmans, 1999.

———. "*Similis et Dissimilis*: Gauging Our Expectations of the Early Fathers." In *Ancient Faith for the Church's Future*, edited by Mark Husbands and Jeffrey P. Greenman, 69-89. Downers Grove, IL: IVP Academic, 2008.

Zimmermann, Jens. *Hermeneutics: A Very Short Introduction*. Oxford: Oxford University Press, 2015.

3

WHAT DOES
"LITERAL" MEAN?

PATRISTIC EXEGESIS IN CONTEXT

Today we make a distinction between "literal" and "spiritual" — which includes moral, allegorical, and anagogical — readings of Scripture. We tend to assume that this is the way it has always been in Christianity. But in the early church, the literal sense of Scripture had a complexity that tends to obscure both the distinction as well as the definition we give to the literal. The actual exegesis of the church fathers is strongly goal-oriented and complicates our desire to categorize and limit it.[1]

For example, Theophilus of Antioch's style of biblical interpretation is often called literal by some patristic scholars.[2] But even though he offers a literal interpretation of the first six days of creation, he is not averse to using typology and allegory and connecting it to the larger Christian story.[3] When he discusses the creation of the luminaries on

---

[1]Paul Blowers, *Drama of the Divine Economy: Creator and Creation in Early Christian Theology and Piety*, Oxford Early Christian Studies (Oxford: Oxford University Press, 2012), 105.
[2]For example, Robert M. Grant, "Theophilus of Antioch to Autolycus," *Harvard Theological Review* 40, no. 4 (1947): 254-55.
[3]Blowers, *Drama of the Divine Economy*, 105.

the fourth day, he claims that they contain "a pattern and type of a great mystery."[4] Because the sun is greater in brightness and power than the moon, the sun is a "type" of God and the moon a "type" of man. Further, the three days preceding the creation of the luminaries are "types" of the triad of God, Logos, and Sophia, and the fourth day is a "type" of man.[5] Irenaeus appears to understand the sequence of days temporally. But, as Blower states, "It becomes clear that these 'days,' over and beyond any simple chronology, all belong to a larger rhythm of providentially guided outcomes and are all part of the mystery of salvation in Christ."[6]

When the Fathers are appropriated in an issue like the creation/ evolution controversy, great care needs to be taken to avoid reading our own understandings and distinctions into the ancients' writings. Unfortunately, as we have seen in the previous chapter, this is not what is happening in some circles. The necessity of a foundational chapter on patristic exegesis of the Bible should, therefore, be glaringly apparent.

In an article appealing to the Fathers as important theological resources, James Mook seeks to ground his call to read the days of Genesis as twenty-four-hour periods of time in the church fathers.[7] Since his main goal is to "counter some of the misreadings of the fathers," he proceeds to offer what he believes is a contextual understanding of their interpretive framework.[8] Mook claims that there was a "tension" among the church fathers between "allegorists and literal interpreters."[9] Although Mook does not specifically mention it, the claim of a tension in the early church is an issue that has received much attention in

---

[4]Theophilus of Antioch, *AD Autolycum* 2.15 (Grant, 51). All quotations of *AD Autolycum* are from Theophilus of Antioch, *Ad Autolycum*, trans. Robert M. Grant, OECT (Oxford: Clarendon Press, 1970).

[5]Ibid. (Grant, 53).

[6]Blowers, *Drama of the Divine Economy*, 106.

[7]James R. Mook, "The Church Fathers on Genesis, the Flood, and the Age of the Earth," in *Coming to Grips with Genesis: Biblical Authority and the Age of the Earth*, ed. Terry Mortenson and Thane H. Ury (Green Forest, AZ: Master Books, 2008), 23-52.

[8]Ibid., 24-25.

[9]Ibid., 29.

patristic studies. It is commonly addressed as the difference between the Alexandrian (spiritual/allegorical) and the Antiochene (literal) approach to biblical interpretation. Unfortunately, it is also the source of much confusion. The simple delineation of the church fathers into allegorists or literalists not only oversimplifies the context but also glosses over some very important issues that have bearing on mustering patristic support for a literal (as understood by Mook) reading of Genesis. In more explicit terms, an examination of the Alexandrian and Antiochene ways of reading Scripture reveals the inadequacy, at best, of equating the Fathers' "literal" reading with the "literal" reading advocated by groups like Answers in Genesis (AiG), Institute for Creation Research (ICR), and Creation Ministries International (CMI).

Mook betrays certain modern assumptions about biblical interpretation in his understanding of the church fathers. There is a certain degree of accuracy in pointing to a tension among the Fathers, but there are too many assumptions made about the relation of patristic exegetical expectations and our modern expectations. Did the Fathers really practice the historical-grammatical method of exegesis that led them to interpret Genesis as "actual history," as Mook and others claim? Addressing these issues requires both an understanding of modern expectations and a proper contextual understanding of the tension between Alexandrian and Antiochene forms of exegesis.

GRAMMATICAL-HISTORICAL METHOD

AiG and CMI are organizations committed to the Bible as the primary source for Christian faith and life. This commitment to taking the Bible seriously is well established in evangelicalism, a movement to which both organizations claim allegiance.[10] Since the Bible is primary, proper study of it is of utmost importance and is expected to follow

---

[10]For a basic understanding of the multifaceted nature of evangelicalism, see Craig D. Allert, "Evangelical Identities: Streams of Confluence and Historical Theology in Evangelicalism," *Canadian Evangelical Review* 37 (2010): 1-20.

certain rules in order to produce proper interpretation. This is not to say, however, that all evangelicals are in agreement on what these rules are or how to apply them in biblical interpretation.

Historically speaking, evangelicalism has been influenced by something called the historical-critical method (or historical criticism, hereafter denoted as HC). The influence has been both negative and positive. For example, negatively speaking, in 1978 Eta Linnemann famously rejected the historical-critical methods she learned under her teacher Rudolf Bultmann because she concluded that its presuppositions were incompatible with evangelical Christianity.[11] Yet, positively speaking, some evangelicals claim there is a legitimacy to at least *some* of the questions asked by HC.[12] Advocates here would argue that evangelicals can accept modified forms of HC provided it is "chastened" and subordinated to the demands of faith.[13] In other words, some evangelicals claim to judiciously use historical-critical methods, while others explicitly reject them.

Historical criticism is properly understood as a grouping of methods rather than one specific method, and forms of it are taught at many evangelical Bible colleges and seminaries throughout North America. When applied to biblical books, HC is a comprehensive term that designates several techniques used to discover the historical situation, sources behind the writings, literary style and relationships, date, authorship, approach to composition, destination, and recipients.[14] Generally

---

[11]Linnemann states that she "repented of her perverse theological training." Eta Linnemann, *Historical Criticism of the Bible: Methodology or Ideology? Reflections of a Bultmannian Turned Evangelical*, trans. Robert W. Yarbrough (Grand Rapids: Baker, 1990). See also Robert W. Yarbrough, "Eta Linnemann: Friend or Foe of Scholarship?," *The Master's Seminary Journal* 8, no. 2 (1997): 163-89.

[12]Gregory Dawes, "'A Certain Similarity to the Devil': Historical Criticism and Christian Faith," in *Interdisciplinary Perspectives on the Authority of Scripture: Historical, Biblical, and Theoretical Perspectives*, ed. Carlos R. Bovell (Eugene, OR: Pickwick, 2011), 356.

[13]S. A. Cummins, "The Theological Interpretation of Scripture: Recent Contributions by Stephen E. Fowl, Christopher R. Seitz and Francis Watson," *Currents in Biblical Research* 2 (2004): 182, 184. See also Dawes, "Certain Similarity to the Devil," 356.

[14]David Dockery, *Christian Scripture: An Evangelical Perspective on Inspiration, Authority, and Interpretation* (Nashville: Broadman & Holman, 1995), 153.

speaking, HC seeks to answer one overarching question – To what historical circumstances does a given text refer, and out of what historical circumstances did it emerge?[15]

As indicated above, evangelicals have had an uneasy relationship with HC. On the one hand, because of its strong connections to Enlightenment rationalism and antisupernaturalism, some tend to be justifiably wary of it. On the other hand, its insistence on grounding proper interpretation in the original author's intent has dominated evangelical thought and practice in biblical interpretation. The attitude of some evangelicals toward a chastened HC is summarized in Alan Johnson's article "The Historical-Critical Method: Egyptian Gold or Pagan Precipice?"[16] In the article, Johnson applies a comment made by Augustine of Hippo in *On Christian Doctrine* to the evangelical use of HC. Augustine writes:

> Just as the Egyptians had not only idols . . . so also they had vases and ornaments of gold and silver and clothing which the Israelites took with them when they fled. . . . In the same way all the teachings of the pagans contain not only simulated and superstitious imaginings . . . but also liberal disciplines more suited to the uses of truth. . . . When the Christian separates himself from their miserable society, he should take this treasure with him for the just use of teaching the gospel.[17]

The argument is that it is not necessarily the historical-critical *method* that is bad but rather the *alien presuppositions* to which certain scholars subject it. The method can be highly serviceable for the evangelical to uncover the past when it is freed from the arbitrary assumptions of the critics.[18]

---

[15]Richard E. Burnett, "Historical Criticism," in *The Dictionary of Theological Interpretation of the Bible*, ed. Kevin J. Vanhoozer (Grand Rapids: Baker Academic, 2005), 290.

[16]Alan F. Johnson, "The Historical-Critical Method: Egyptian Gold or Pagan Precipice?," *Journal of the Evangelical Theological Society* 26 (1983): 3.

[17]Augustine, *On Christian Doctrine*, 2.40.60, as cited by Johnson.

[18]Grant R. Osborne, "Historical Criticism and the Evangelical," *Journal of the Evangelical Theological Society* 42 (June 1999): 209. See also William W. Klein, Craig L. Blomberg, and Robert L. Hubbard Jr., *Introduction to Biblical Interpretation* (Dallas, TX: Word, 1993), 17.

The reason for the method's serviceability is the strongly held belief that the proper meaning of the text lay in the original meaning and intent of the author. Despite the disagreements evangelicals may have with some of Benjamin Jowett's arguments in a seminal article advocating HC, many tend to agree that "Scripture has one meaning – the meaning that it had in the mind of the Prophet or Evangelist who first uttered or wrote, to the hearers who first received it."[19] Later in the same article Jowett adds, "The true use of interpretation is to get rid of interpretation, and leave us alone in the company of the author."[20] The foundational assumption is that proper interpretation is found in a method aimed primarily at historical reconstruction of the text – that the most primitive meaning of the text is the only valid meaning and that the historical-critical method is the only key that can unlock it.[21]

AiG and CMI self-identify among the evangelicals who explicitly reject HC, which is seen by these groups as "a product of philosophical ideologies that are inherently hostile to the biblical text."[22] F. David Farnell, The Master's Seminary professor of New Testament, wrote an article expanding on this general rejection with more specific rationale for the rejection of HC.[23] Farnell and the other editors of the book in which this article appears advocate an approach to biblical interpretation called the grammatical-historical (GH) method, which is also the approach advocated by AiG and CMI. Farnell emphasizes the necessity of not confusing GH with HC (they must be kept separate) and lists four reasons why advocates of the former reject the latter.

---

[19]Benjamin Jowett, "On the Interpretation of Scripture," *Essays and Reviews*, 7th ed. (London: Longman, Green, Longman and Roberts, 1861), 378. Regarding the disagreements with Jowett, I refer specifically to his claim that the Bible should be read like any other book.

[20]Ibid., 384.

[21]David C. Steinmetz, "The Superiority of Pre-Critical Exegesis," in *The Theological Interpretation of Scripture: Classic and Contemporary Readings*, ed. Stephen E. Fowl (Malden, MA: Blackwell, 1997), 27.

[22]F. David Farnell, "Grammatical-Historical Versus Historical-Critical," in *Basics of Biblical Criticism: Harmful or Helpful*, 2nd ed., ed. F. David Farnell, Thomas Howe, Thomas Marshall, Benjamin Cocar, Dianna Newman, and Edward D. Andrews (Cambridge, OH: Christian Publishing House, 2016), 39.

[23]Farnell, "Grammatical-Historical Versus Historical-Critical," 37-41.

First, the roots of HC lay in the deism and rationalism of the Enlightenment. Even proponents of the method readily admit that "historical method is the child of the Enlightenment."[24] Second, HC assumes Ernst Troeltsch's (1865–1923) ideological principles. Troeltsch was a German philosopher and theologian who was one of the best-known defenders of HC and, in the eyes of GH practitioners, a theological liberal. Interestingly, Troeltsch himself recognized some of the challenges HC posed. He actually stated that "from a strictly orthodox standpoint" HC "seems to bear a certain similarity to the devil."[25] Still, Troeltsch believed that this approach was not necessarily contrary to the Christian faith. He argued that there are three principles that underlie the historical method. Farnell summarizes these principles in this way:

> (a) the principle of criticism or methodological doubt – history achieves only probability, nothing can be known with any certainty;
>
> (b) the principle of analogy (somewhat like the modern idea of uniformitarianism) that present experience becomes the criteria of probability in the past (hence, if no supernatural events occur today; then, they do not occur in the past either);
>
> (c) correlation or mutual interdependence that postulates a closed continuum of cause and effect with no outside divine intervention.[26]

The third reason HC is rejected by advocates of the GH method is because it is believed to pursue a deductive rather than an inductive approach. Farnell asserts that the deductive approach assumes an interpretation that Scripture is then forced into rather than reading the meaning out of Scripture – this is often expressed as the difference between eisegesis (deductive) and exegesis (inductive). Fourth, HC practices "dependency hypotheses." Farnell fails to explain what he means here,

---

[24]Edgar Krentz, *The Historical-Critical Method* (Philadelphia: Fortress Press, 1975), 55.

[25]Ernst Troeltsch, "Historical and Dogmatic Method in Theology (1898)," in *Religion in History – Ernst Troeltsch*, trans. J. L. Adams and W. F. Bense (Edinburgh: T&T Clark, 1991), 16, as cited in Dawes, "Certain Similarity to the Devil," 357.

[26]Farnell, "Grammatical-Historical Versus Historical-Critical," 37-38.

but we can confidently assume that he is referring to the theory of literary dependence, which claims that a Gospel writer composed his Gospel by copying from, and thus modifying, a written exemplar of different Gospels or sources.[27] Because dependency theories arose in the modern period with its roots in skepticism of the biblical record, it is argued that acceptance of dependence theories automatically rejects the doctrine of the inerrancy of Scripture.

Advocates of the GH method lament the unfortunate choice that evangelicals are forced to make between it and HC because they "oppose one another in dramatic ways."[28] Those lamentations are reflected in articles on biblical interpretation found on the AiG and CMI websites. That is, their approach to biblical interpretation (GH) is always set in opposition to HC. In what follows, I will mainly use two articles from the AiG website to explain how AiG and CMI interpret the Bible, although these will be supplemented in places by others.[29]

Foundational to GH is the verbal inspiration and inerrancy of the Bible. It is literally God's word – his message – to humanity, and because of this "we can have perfect confidence that God is capable of accurately relaying His Word to us in a way that we can understand."[30] They do, however, understand that while belief in the inspiration of the Bible is essential, it would be self-defeating if correct interpretation did not accompany any assertion of inspiration: "There would be little value in being able to say, 'These are the words of God,' if we then interpret them in a way God never intended."[31] Thus the necessity of correct interpretation.

---

[27]See Robert L. Thomas and F. David Farnell, *The Jesus Crisis: The Inroads of Historical Criticism into Evangelical Scholarship* (Grand Rapids: Kregel, 1998).

[28]Robert L. Thomas, "Current Hermeneutical Trends: Toward Explanation or Obfuscation?," *Journal of the Evangelical Theological Society* 39, no. 2 (1996): 241.

[29]Tim Chaffey, "How Should We Interpret the Bible, Part 1: Principles for Understanding God's Word," February 22, 2011, https://answersingenesis.org/hermeneutics/how-we-interpret-the-bible -principles-for-understanding/; and Brian H. Edwards, "Unlocking the Truth of Scripture," October 1, 2007, https://answersingenesis.org/hermeneutics/unlocking-the-truth-of-scripture/.

[30]Chaffey, "How Should We Interpret."

[31]Edwards, "Unlocking the Truth."

God intended to communicate something to us in the Bible, and our responsibility is to interpret it correctly. This conviction is often expressed in terms of an analogy. For example, Brian Edwards states that the Bible is a "treasure box." In order to get to the treasure within, a key must be used by Christians to unlock it. That key is the grammatical-historical method. Similarly, In *How to Study the Bible*, The Master's Seminary president John MacArthur compares Jesus' words in John 8:31-32 to the man who works on a math problem and finds the answer, or the scientist in the lab pouring solutions.[32] The scientist "stays with it until he says, 'Eureka, I found it!' – then he's free."[33] AiG president Ken Ham prefers the analogy of a spiritual virus that is rampant in many seminaries and Bible colleges and has caused many to interpret the Bible incorrectly. The antidote to this virus is "a 'spiritual vaccine' that teaches a way to think that enables people to 'interpret' God's Word correctly and believe and understand this special revelation of absolute truth."[34] Later in the same article, Ham employs the analogy of two keys to illustrate the difference between eisegesis and exegesis. Eisegesis is the wrong key and will not unlock the correct meaning of Scripture, but the correct key, exegesis, will – provided it is done in the GH manner.

In order to avoid reading one's ideas into Scripture (eisegesis), the interpreter's primary goal in approaching any passage in the Bible is to determine the author's intended meaning, which includes how the original hearers would have understood the author. Since God communicated his message to the human authors, the interpreter must be intent on finding it. We must "understand the text as God and the human writers intended" because a "given document means what the

---

[32]Jn 8:31-32: "If you continue in my word . . . you will know the truth, and the truth will make you free."

[33]John F. MacArthur, *How to Study the Bible* (Chicago: Moody Publishers, 2009), loc. 262-63, Kindle edition.

[34]Ken Ham, "Eisegesis: A Genesis Virus," June 1, 2002, https://answersingenesis.org/hermeneutics /eisegesis/.

author intended it to mean."[35] CMI speaker Calvin Smith explains that the GH method attempts "to interpret the text in the way the author intended."[36] Because basic communication is God's intention in giving us these written texts in the Bible, it is self-evident that the interpreter follow the standard rules of grammar and interpretation in exegesis.

The foundational presuppositions to biblical interpretation are a necessary prelude to more specific principles of the GH method of biblical interpretation that must be followed. AiG's Bodie Hodge and CMI's Calvin Smith each claim that the Bible gives us "principles of interpretation."[37] In support of this claim they both cite 2 Corinthians 4:2 and Proverbs 8:8-9:[38]

> Rather, we have renounced secret and shameful ways; we do not use deception, nor do we distort the word of God. On the contrary, by setting forth the truth plainly we commend ourselves to everyone's conscience in the sight of God. (2 Corinthians 4:2 NIV)

> All the words of my mouth are righteous; there is nothing twisted or crooked in them. They are all straight to one who understands and right to those who find knowledge. (Proverbs 8:8-9 NRSV)

It is difficult to see how these verses indicate "principles for biblical interpretation" as the authors claim. They, of course, are not principles. Rather, they are cited to bolster their insistence that the Bible is to be read in a *plain* or *straightforward* manner. This is what they mean by literal interpretation.

Tim Chaffey explains that, in order to read the Bible plainly, or literally, six principles must be followed to guide the interpreter in accurately interpreting God's Word. The principles "do not comprise an

---

[35]Edwards, "Unlocking the Truth"; and Chaffey, "How Should We Interpret."

[36]Calvin Smith, "Is There a Universal Way Christians Should Interpret the Bible?," February 5, 2013, http://creation.com/is-there-a-universal-way-christians-should-interpret-the-bible.

[37]Bodie Hodge, "Why Do You Take the Bible Literally?," January 13, 2006, https://answersingenesis .org/bible-questions/why-do-you-take-the-bible-literally/; and Smith, "Is There a Universal Way?"

[38]These verses are cited here as they appear in Hodge and Smith. Neither indicates that the verses are from different translations, nor why they have chosen to cite different translations.

exhaustive list but are some of the major concepts found in the majority of books on interpretation."[39] Since I could not find any articles on the websites that detailed the principles in the same order, I will follow the order laid out by Chaffey and supplement that order with others.

Chaffey reminds his readers that the primary goal in biblical interpretation is to determine the author's intended meaning. In order to reach that goal, readers are encouraged to follow principles that are "derived from God's Word." Nowhere, however, does Chaffey, or other advocates of GH, detail where these are found or how they are derived from the Bible (more on this below).

Principle one: Observe the text. The first principle is to carefully observe the text. This is a very basic yet necessary step. Too many people, argues Chaffey, have made mistakes based on what they *think* the text states rather than what it does actually state. Here grammatical constructions and syntax must be carefully observed and noted. Edwards discusses these kinds of things under the heading "What Is the Plain Meaning?"[40] Here the exegete is to look for the grammatical sense and meaning of the words. He exhorts the reader not to look for "some mysterious, hidden meaning" if the plain sense makes good sense. Support for this is said to come from Jesus himself when he said "Have you not read?" For Edwards, this is affirmation that Jesus "obviously thought that scripture is basically clear."[41] Because Scripture is clear, it should follow the basic rules of grammar.

Principle two: Context. The second principle of GH interpretation is that context is key.[42] According to Chaffey, this is the most agreed-on

---

[39]Chaffey, "How Should We Interpret." Books on interpretation recommended by Chaffey and other AiG and CMI authors include MacArthur, *How to Study the Bible*; and Milton S. Terry, *Biblical Hermeneutics: A Treatise on the Interpretation of the Old and New Testaments* (Grand Rapids: Zondervan, 1974).

[40]Edwards, "Unlocking the Truth."

[41]Ibid. Edwards does not cite the passages where Jesus makes this statement. New Testament translations have several occurrences of this statement in one form or another ("Have you not read?"; "Haven't you read?"; "Have you not even read?"): Mt 12:3, 5; 19:4; 21:16, 42; 22:31; Mk 12:10, 26; Lk 6:3.

[42]Chaffey, "How Should We Interpret."

principle of biblical interpretation. He uses Webster's dictionary to define context as "the parts of a discourse that surround a word or a passage and can throw light on its meaning." Failure to take account of context often leads to eisegesis, where interpreters make the Bible "say" what they want it to say. It is important for the interpreter to be aware of the context of particular passages and books as well as how individual books fit with the whole of Scripture. For Edwards, this means identifying the audience, topic, and theme of the book in which the passage is found.[43] MacArthur writes about a "historical principle" that is concerned with what the text meant to the people to whom it was written or spoken. Understanding the historical context is essential; otherwise, "you'll never really understand what's in the writer's heart."[44]

Context also includes recognizing how passages fit in with the flow of history. Chaffey explains that it "makes a huge difference in determining the writer's intent if we note whether the passage was pre-Fall, pre-Flood, pre-Mosaic Law, after the Babylonian Exile, during Christ's earthly ministry, after His Resurrection, or after Pentecost."[45] Edwards concurs and adds that the historical context is often found in the Bible itself. However, readers should consult good Bible commentaries or Bible encyclopedias for help.

Principle three: Clarity of Scripture. We have already discussed this above. The Bible is God's communication to humanity. Since God desired to communicate to us, we should expect that he wanted us to understand. This is why the Bible follows the normal modes of communication and why we expect it to follow the normal rules of grammar.

Principle four: Compare Scripture with Scripture. Since Scripture comes from God, there will be no contradiction in it. This is sometimes called the "analogy of faith" or the "analogy of Scripture." While the

---

[43]Edwards, "Unlocking the Truth."
[44]MacArthur, *How to Study the Bible*, loc. 1235-36.
[45]Chaffey, "How Should We Interpret."

previous principle emphasized the clarity of Scripture, this one gives a significant qualification to that clarity. Since "not all Bible passages are equally clear," we should use clear passages to shed light on difficult ones.[46] In fact, there are a number of "obscure" verses in Scripture where we wish the biblical author had provided more detail. In those cases, it may be that we simply cannot come to a firm conclusion about its meaning. Still, the comparison principle provides interpreters with a system of checks and balances to help keep us on the right track. In fact, Edwards claims that it can provide us with "great certainty" about any given interpretation, such that "we can be confident in the accuracy of our interpretation."[47] Professor Emeritus Robert L. Thomas of The Master's Seminary even claims that the Christian interpreter is capable of neutral objectivity in interpreting: "Neutral objectivity originates with the creator of all things and is available through the illumination of the Holy Spirit."[48] Thus, human limitations do not distort the process of divine communication with humanity. The history of the GH method of interpretation is evidence of this.[49]

Principle five: Classification of texts. Classification is a reference to understanding the literary style or genre of passages in interpretation. Since the Bible contains numerous types of literature, the reader needs to be aware of what sort of literature is being interpreted. The Bible is described as having four main kinds of literature: history, poetry, prophecy, and epistles (teaching letters). The interpreter needs to be aware of what to expect in each particular kind. Since Genesis is written as history, we should expect to interpret it in a straightforward manner as history. We should "remember that its purpose is to describe things that actually happened."[50] However, books can contain more than one style. A historical book like Exodus contains a song

---

[46]Edwards, "Unlocking the Truth."
[47]Ibid.
[48]Thomas, "Current Hermeneutical Trends," 254.
[49]Ibid., 255.
[50]Edwards, "Unlocking the Truth."

written in poetic language in chapter fifteen and, therefore, employs figurative language.[51]

Principle six: Take account of the church's past. The final principle of the GH method of interpretation according to Chaffey is the church's historical view. Even though the interpreter's conclusions must be based solidly on the Bible, it is important to know how those who have gone before us have interpreted a passage in question. This is an important resource because there are many who have spent long hours studying the Bible and discussing doctrines. We should take advantage of that resource. It would be a significant red flag if one reached a conclusion today that no one else in history reached. It does not mean one would necessarily be wrong, but it should signal a review of the interpretation in question.

Supplemental principle: Application. Application is also an important aspect, albeit not really a part of the process as much as it is a conclusion of it. The Bible was not written merely to give intellectual stimulation – accumulation of knowledge is not the goal. Rather, we study the Bible and interpret it accurately "so that we may know God better, know what he expects from us, and know how we can live in a way that pleases him."[52] The proper way to do this is to find, first and foremost, the author's intended meaning. It cannot mean something to us that it did not mean to the original readers or hearers.

EVALUATING THE CLAIMS

Organizations like AiG and CMI appropriate the church fathers as advocates of a nascent creation science position. But even more foundational than this is the claim that the authors of the New Testament and "theologians since the Fathers" were practitioners of the GH method of interpretation. Related to this overarching claim are two specific assertions – one regarding Jesus' words on the "clarity" of Scripture and the other on the inappropriateness of allegory.[53]

---

[51]Chaffey, "How Should We Interpret."
[52]Ibid.
[53]Of course, these are not the only issues that could be examined. I will, for example, forgo any

The inappropriateness of allegory. Not much is said about allegory in these works on proper method in biblical interpretation. What is said, however, is entirely negative. In his popular book *How to Study the Bible*, John MacArthur discusses what he calls "errors in interpretation" in relation to the clarity of Scripture. He recalls the first sermon he ever preached and describes it as "horrible": "My text was 'And the angel rolled the stone away.' My sermon was 'Rolling Away Stones in Your Life.' I talked about the stone of doubt, the stone of fear, and the stone of anger."[54]

He bemoans his sermon because the verse is about a "real stone," and by allegorizing he had betrayed the historical referent. Similarly, a sermon he once heard on Acts 27:29[55] that referred to the anchors as hope, faith, and so on caused MacArthur to insist that they "were not anchors of anything but metal." Thus, the caution offered by MacArthur is: "You must not spiritualize" in biblical interpretation. In doing so, the Bible is made into a fairy-tale book from which we can get "all kinds of crazy interpretations."

Later, when discussing the "literal principle," MacArthur warns about those who claim a secret meaning in the Bible and then employ an allegorical method to get this hidden meaning. His response to those who practice this is less than complimentary: "Do you know what that is? Nobody knows! They make it up. Don't do that – interpret Scripture in its literal sense."[56] Similarly, Calvin Smith denigrates allegory for its lack of connection to the historical event. Allegory is "imposed on the text" and is devoid of "rules of interpretation."[57]

The caricatures of allegory presented by MacArthur and Smith raise the question – Is this an accurate portrayal of committed Christians, like the Fathers, who interpreted allegorically? Should we not take the

---

extended analysis of the claims of neutral objectivity in biblical interpretation.
[54]MacArthur, *How to Study the Bible*, loc. 1163-64.
[55]Acts 27:29 (KJV): "they cast four anchors . . . and wished for the day."
[56]MacArthur, *How to Study the Bible*, loc. 1232-33.
[57]Smith, "Is There a Universal Way?"

advice of Chaffey, who claims that GH practitioners "are informed by the work of others who have spent long hours studying God's Word"?[58] Has this happened when it comes to understanding allegorical/spiritual interpretation?

The failure to accurately represent allegory in the AiG and CMI literature is a problem. For all the talk about proper interpretation, context, and plain meaning, it really is shameful that the accurate representation of viewpoints with which they disagree is dismissed out of hand. Later in this chapter a contextual understanding of allegorical interpretation in the early church will be presented. We will see that the claims made above about allegory are not situated in the context in which allegorical interpretation was practiced. The idea that allegorical readings are made up or not guided by any principles or rules is simply an inaccurate assertion.

"Have you not read?" The common expression of Jesus, "Have you not read?" is claimed to be an explicit statement about the clarity of Scripture. Brian Edwards advises that the plain meaning of Scripture is the aim of the interpreter, not a hidden meaning. No other sense should be sought if the literal (plain) makes good sense. For Edwards, when Jesus made these statements, "obviously he thought that Scripture is basically clear."[59] Chaffey agrees because when Jesus made these statements he would follow them with a quote from the Old Testament. Thus, "by these sayings, He indicated that the Scriptures are basically clear."[60]

Is this really what Jesus' statements indicate? Is this their plain reading? Jesus makes the statement nine times in the Gospels.[61] Six of the nine occurrences appear in the Gospel of Matthew as a question posed by Jesus to groups of religious leaders. The passages are very good examples of where grammatical construction rubs against context,

---

[58]Chaffey, "How We Should Interpret."
[59]Edwards, "Unlocking the Truth."
[60]Chaffey, "How We Should Interpret."
[61]Mt 12:3, 5; 19:4; 21:16, 42; 22:31; Mk 12:10, 26; Lk 6:3.

thus indicating that communication is not always necessarily as "plain" as our authors insist.[62] Jesus' statement actually cuts two ways—as both a compliment and an insult. On the one hand, the grammatical construction of the phrase "Have you not read?" shows that Jesus appeals to common ground with his opponents.[63] It is similar to saying something like "Of course you've read the Scripture, haven't you?" He expects a high degree of biblical literacy among his opponents, which includes Pharisees, scribes, and chief priests. On the other hand, there is a polemical context at work that indicates an inability to properly understand Scripture. It is similar to saying something like "While I'm sure you've read the Scriptures, you don't act like you've really understood them or grasped their full significance." It is helpful to take a closer look at an extended passage in which the statement occurs twice—Matthew 12:1-8:[64]

> At that time Jesus went through the grainfields on the sabbath; his disciples were hungry, and they began to pluck heads of grain and to eat. When the Pharisees saw it, they said to him, "Look, your disciples are doing what is not lawful to do on the sabbath." He said to them, "Have you not read what David did when he and his companions were hungry? He entered the house of God and ate the bread of the Presence, which it was not lawful for him or his companions to eat, but only for the priests. Or have you not read in the law that on the sabbath the priests in the temple break the sabbath and yet are guiltless? I tell you, something greater than the temple is here. But if you had known what this means, 'I desire mercy and not sacrifice,' you would not have condemned the guiltless. For the Son of Man is lord of the sabbath."

The narrative here has three actors—Jesus, his disciples, and the Pharisees. The Pharisees, who are responsible for correcting infractions

---

[62]What follows draws on Scott F. Spencer, "Scripture, Hermeneutics, and Matthew's Jesus," *Interpretation* 64, no. 4 (2010): 368.

[63]The interrogative construction "Have you not (*ou/ouk*) never (*oudepote*)?" plus an indicative verb usually expects an answer in the affirmative.

[64]What follows draws on Paul S. Minear, "On Seeing the Good News," *Theology Today* 55, no. 2 (1998): 171-73.

of the law, condemn the disciples for their transgression. But Jesus turns the tables, condemning the Pharisees and exonerating the disciples. What is the guilt of the Pharisees here? "They did not recognize in Jesus and his disciples the authentic successors of David and his companions."[65] They know the law, for they have indeed read it. But Jesus tells them "something greater" is here that the Pharisees do not see and understand because they are, in fact, *not* reading. Jesus thus quotes Hosea 6:6 to indicate the presence of something greater than the need for sacrificial offerings, something greater than the Pharisees' sabbath.[66]

Jesus' statement is not used by the Gospel of Matthew as a defense for the clarity of Scripture. His opponents read the Scriptures and had intimate knowledge of them. Yet they did not understand them; it was actually unclear to them because they were reading it as it was meant by the original authors! Is Jesus not calling here for a deeper knowledge based on his incarnation? Something greater is here, and this requires a reading that goes beyond its original intent – a reading that is based on an understanding of Jesus as the fulfillment of the law. His opponents need to understand this in order to properly understand Scripture. They read but do not really know. For them, the Scriptures are not clear because they miss the central significance of Christ.

The irony, of course, is that Edwards and Chaffey have made a claim about these statements without taking the advice they offer in their own article. One could argue that, contrary to their adamant assertions to read out of a passage (exegete), they have fallen prey to the "virus" of eisegesis. Nowhere do they justify their assertion about the words of Jesus. They assert and move on. But there is certainly a contextual justification for reading this question of Jesus as a call to deeper readings.

Further, Chaffey has failed to take his own advice and consider what others in Christian history have to say about these verses. Understanding

---

[65]Ibid., 172.
[66]Hos 6:6: "I desire steadfast love and not sacrifice."

these words as calls to deeper readings is how the earliest interpreters comprehended them. For example, the Antiochene church father John Chrysostom (349–407) addresses Jesus' statement in Homily 39 on Matthew 12:1.[67] In the homily, Chrysostom points out that Jesus refers to David because of the great glory of that "prophet."[68] Why, Chrysostom asks, speak about David and his greatness when there was a transgression of the law by the disciples? The disciples are acquitted of the transgression because "he who is greater is found to have done the same."[69]

For Chrysostom, this indicates that Christ, who is also greater, brings a greater sabbath. Both David's eating the bread of the presence and the temple priests' breaking the sabbath set a precedent for others. Jesus defends the disciples by appealing to the actions of the priests in the temple. Chrysostom recognizes a possible objection – "But [the disciples] are not priests." His response appeals to his theme of greatness: "Nay, they are greater than priests. For the Lord of the temple Himself is here: the truth, not the type. Wherefore He said also, 'But I say unto you, that in this place is one greater than the temple.'"[70] For Chrysostom, something more significant is happening than the mere narrative of a story and an appeal to the clarity of Scripture. It is significant for him that the Pharisees made no reply to Jesus. This is because "the salvation of men was not their object."[71]

Chrysostom understands this narrative as somehow being about salvation. What I have described as David and the temple priests setting precedents, Chrysostom calls types. Jesus rebukes the Pharisees by quoting Hosea 6:6 ("I desire steadfast love and not sacrifice"). If they had known what the passage *means*, they would not have condemned the guiltless. For Chrysostom the law referred to Christ. In this passage

---

[67]John Chrysostom, Homily 39 (NPNF[1] 10.255-58).
[68]Ibid. (NPNF[1] 10.255).
[69]Ibid. (NPNF[1] 10.256).
[70]Ibid., 39.2 (NPNF[1] 10.256-57).
[71]Ibid., 39.2 (NPNF[1] 10.257).

in Matthew it is on Jesus' own authority that he makes these claims. Yet "this too is out of the law."[72] The law spoke of Christ through types.

The final words of the Matthew narrative (12:8) relay Jesus' statement, "For the Son of Man is lord of the sabbath." Chrysostom points out that the Gospel of Mark expresses it somewhat differently: "The Sabbath was made for man, not man for the Sabbath."[73] The sabbath had many benefits "at first." One of the greatest was that "it trained them by degrees to abstain from wickedness, and disposed them to regard the things of the Spirit."[74] The appeal here is to the exodus via Ezekiel 20, where God continually shows grace to Israel in the midst of their continual rebellion—they profaned God's sabbaths, but God did not destroy them in the wilderness. God,

> in the very act of giving the law of the Sabbath, did even therein darkly signify that He will have them refrain from evil works only, by the saying, 'ye must do no work, except what shall be done for your life' [Ex 12:16]. And in the temple too all went on, with more diligence and double toil [Num 28:9-10]. Thus even by the very shadow He was secretly opening them the truth.[75]

Christ did not put an end to the sabbath, which was actually profitable. Rather, "he greatly enhanced it. For it was time for them to be trained in all things by the higher rules." Now that Christ is come, the true understanding of the law is revealed. Indeed, "Why is any Sabbath required, by him who is always keeping the feast, whose conversation is in heaven?"[76]

Chrysostom exemplifies a standard understanding of Jesus' words among the church fathers. It is a call to comprehend the momentous change the incarnation of Christ brought and the need to view him as "greatly enhancing" the law. In other words, because of Christ, the

---

[72]Ibid.
[73]Mk 2:27, as quoted by Chrysostom, ibid., 39.3 (NPNF[1] 10.257).
[74]Chrysostom, Homily 39.3 (NPNF[1] 10.257).
[75]Ibid. (NPNF[1] 10.257).
[76]Ibid.

Pharisees need to read differently, seeking Christ in the Old Testament, not
the author's intended meaning. What is this other than a hidden meaning
that MacArthur, Chaffey, Edwards, and Smith explicitly reject because it is
imposed on the text and allows the interpreter to make things up? Chrysostom states that the claims made by Jesus were based on his own authority. But he is also sure to point out that they were also "from the Law."[77]
Even though hidden, this is what the text really says. This leads us to what
is perhaps the most blatant inaccuracy these authors assert.

NEW TESTAMENT AUTHORS AND
FATHERS AS PRACTITIONERS OF THE
GRAMMATICAL-HISTORICAL METHOD

Tim Chaffey states, "It can be demonstrated that the New Testament
authors interpreted the Old Testament in this manner [GH]."[78] But the
failure to actually offer support for the claim leaves one wondering
about its accuracy. Chaffey, Hodge, and Smith all claim that the GH
method of biblical interpretation is the same as that practiced by the
New Testament writers and by theologians since the church fathers.
Presumably, the phrase "since the church fathers" is inclusive of them.[79]
The example of John Chrysostom above should already invite some
hesitation about that claim. But before we get to a more detailed look
at the Fathers and their approach to biblical interpretation, we need to
test the claim against the New Testament writers, which Chaffey, Hodge,
and Smith all claim as allies in their approach to hermeneutics.

In many ways the GH method of interpretation as it is described on the
AiG and CMI websites is actually at odds with the way the first interpreters
of Scripture (the Old Testament) approached interpretation. I doubt that
the apostle Paul would be viewed as a very good interpreter of Scripture
if the GH ideal was the standard against which his interpretation was

---

[77]Ibid., 39.2 (NPNF[1] 10.257).
[78]Chaffey, "How Should We Interpret."
[79]Hodge, "Why Do You Take the Bible Literally?"

measured. Perhaps this is why Hodge, when he advocates for GH, omits Paul as one who employed the method.[80] How would Paul's interpretation of certain Old Testament texts be judged in light of the grammatical-historical method?

Galatians 4:21-26.

Tell me, you who desire to be subject to the law, will you not listen to the law? For it is written that Abraham had two sons, one by a slave woman and the other by a free woman. One, the child of the slave, was born according to the flesh; the other, the child of the free woman, was born through the promise. Now this is an allegory: these women are two covenants. One woman, in fact, is Hagar, from Mount Sinai, bearing children for slavery. Now Hagar is Mount Sinai in Arabia and corresponds to the present Jerusalem, for she is in slavery with her children. But the other woman corresponds to the Jerusalem above; she is free, and she is our mother.

The point here is not to offer an apologetic for allegory because of its much debated employ in this verse. Rather, it is to observe how Paul interpreted the narrative of the two sons of Abraham in Genesis. Ishmael, the child of Hagar the slave, was born "according to the flesh," according to ordinary ways of sexual desire, fertility, and insemination.[81] Isaac, the child of Sarah the free woman, was born "through the promise." Paul states that this story is "an allegory" (*allēgoroumena*). His real concern is not the historical narrative here but rather the controversy over whether ritual laws like circumcision should still define the Christian community.[82] For Paul, Hagar is Mount Sinai and represents the earthly Jerusalem. This is where adherence to Jewish law holds sway. Sarah corresponds to the heavenly Jerusalem, where Christians have been set free from bondage to the law.

---

[80]Hodge, in "Why Do You Take the Bible Literally?," states, "Reading the Bible plainly/straightforwardly (taking into account literary style, context, authorship, etc.) is the basis for what is called the *historical-grammatical* method of interpretation which has been used by theologians since the church fathers." Italics in original.

[81]John O'Keefe and R. R. Reno, *Sanctified Vision: An Introduction to Early Christian Interpretation of the Bible* (Baltimore, MD: Johns Hopkins University Press, 2005), 90.

[82]Ibid.

Paul's reading here of the Genesis narrative of the sons of Abraham employs a "method" different from that advocated by our evangelical brothers above. Paul goes beyond the historical ("Hagar is Mount Sinai," etc.) and, in fact, shows no real concern to find the author's intended meaning before he comes to his conclusions. As John O'Keefe and R. R. Reno explain, "In short, Paul develops a reading of the story of Hagar and Sarah and their sons as a map of the divine economy of redemption: first, a carnal covenant of slavery (the precepts delivered by Moses) and then, in fulfillment of the promise, a spiritual covenant of freedom (faith in Christ)."[83] Paul does not limit the text to what can be discovered historically. In fact, he does not even attempt a GH examination and application. He allows later insight that is based on later revelation to influence his inspired reading of the narrative. The context in which Paul interprets is not historical; it is theological.

Origen of Alexandria, in his seventh homily on Genesis, preferred not to comment explicitly on the Hagar and Sarah narrative in Genesis because he believed that Paul had already explained how it was to be understood. In support, he quotes Galatians 4:21-24 and indicates that even though these things happened "in the flesh," they are still to be understood allegorically:

> This, indeed, is what is astonishing in the apostle's understanding, that he called things "allegorical" that are quite obviously done in the flesh. His purpose is that we might learn how to treat other passages, and especially these in which the historical narrative appears to reveal nothing worthy of the divine law.[84]

For Origen, Paul's treatment here stands as a model for how to treat other passages of Scripture. Many church fathers took Origen's understanding of Paul here as axiomatic in their own biblical interpretation.

---

[83]Ibid., 91.

[84]Origen, *Homily on Genesis* 7.2 (FC 71.128-29). All quotations of *Homilies on Genesis* are from Origen, *Homilies on Genesis and Exodus*, trans. Ronald E. Heine, FC 71 (Washington, DC: Catholic University of America Press, 2010).

2 Corinthians 3:12-18.

Since, then, we have such a hope, we act with great boldness, not like Moses, who put a veil over his face to keep the people of Israel from gazing at the end of the glory that was being set aside. But their minds were hardened. Indeed, to this very day, when they hear the reading of the old covenant, that same veil is still there, since only in Christ is it set aside. Indeed, to this very day whenever Moses is read, a veil lies over their minds; but when one turns to the Lord, the veil is removed. Now the Lord is the Spirit, and where the Spirit of the Lord is, there is freedom. And all of us, with unveiled faces, seeing the glory of the Lord as though reflected in a mirror, are being transformed into the same image from one degree of glory to another; for this comes from the Lord, the Spirit.

The reference here is to Exodus 34:33-35, where the narrative describes Moses as placing a veil over his face after speaking with the Lord.[85] For Paul, there is more going on here than what these words mean historically, and, again, he shows no explicit move to find the author's intended meaning. He uses the passage to explain how the true meaning of the Hebrew Scriptures is obscured for the Jews. This, according to Rowan Greer, "is an allegorical warrant for metaphor."[86] But the veil is removed for the Christian because of Christ. Now the Christian can understand the real (spiritual) meaning of Scripture that, in some sense, transcends the historical narrative and the text's original author, readers, and hearers. Once again, the context is not historical, but theological.

1 Corinthians 10:1-11.

I do not want you to be unaware, brothers and sisters, that our ancestors were all under the cloud, and all passed through the sea, and all were baptized into Moses in the cloud and in the sea, and all ate the same spiritual food, and all drank the same spiritual drink. For they drank

---

[85]Ex 34:33-35: "When Moses had finished speaking with them, he put a veil on his face; but whenever Moses went in before the LORD to speak with him, he would take the veil off, until he came out; and when he came out, and told the Israelites what he had been commanded, the Israelites would see the face of Moses, that the skin of his face was shining; and Moses would put the veil on his face again, until he went in to speak with him."

[86]James L. Kugel and Rowan A. Greer, *Early Biblical Interpretation*, Library of Early Christianity (Philadelphia: Westminster Press, 1986), 134.

from the spiritual rock that followed them, and the rock was Christ. Nevertheless, God was not pleased with most of them, and they were struck down in the wilderness. Now these things occurred as examples [*typoi*] for us, so that we might not desire evil as they did. Do not become idolaters as some of them did; as it is written, "The people sat down to eat and drink, and they rose up to play." We must not indulge in sexual immorality as some of them did, and twenty-three thousand fell in a single day. We must not put Christ to the test, as some of them did, and were destroyed by serpents. And do not complain as some of them did, and were destroyed by the destroyer. These things happened to them to serve as an example [*typikōs*], and they were written down to instruct us, on whom the ends of the ages have come.

This passage draws on events found in the book of Exodus and the Israelites' desert wanderings. Paul draws on them, however, not necessarily to reconstruct the historical narrative and find out what the original author meant but to indicate that they were written specifically for "us" — that is, *not* the original audience. "These things," Paul states, "occurred as examples [*typoi* = types, figures]" and "serve as an example [*typikōs*]" *for us* — "they were written down to instruct us." Thus, what the Jews understood as the "crossing of a sea," Paul calls "baptism." The food and drink were "spiritual." The rock was Christ! This leads Robert Wilken to conclude that "Paul's interpretation of Exodus and the wanderings in the desert differs from the 'plain sense' of the text."[87] Once again, the context is theological, not historical.

Origen of Alexandria used this text to indicate his belief that Paul had shown Christians "how the church gathered from the Gentiles ought to interpret the books of the Law."[88] Because Paul dealt with only a few passages out of the mass of Hebrew Scriptures, Origen argues that these kinds of examples should be taken as models to guide Christians in their interpretation of all of Scripture.

---

[87]Robert L. Wilken, *The Spirit of Early Christian Thought: Seeking the Face of God* (New Haven, CT: Yale University Press, 2003), 71.
[88]Origen, *Homily on Exodus* 5.1 (FC 71:275). See Wilken, *Spirit of Early Christian Thought*, 70.

This is very similar to the advice offered by Augustine in *Against Faustus*. He cites the very same 1 Corinthians text with the following conclusion:

> The explanation of one thing is a key to the rest. For if the rock is Christ from its stability, is not the manna Christ, the living bread which came down from heaven, which gives spiritual life to those who truly feed on it? The Israelites died because they received the figure only in its carnal sense. The apostle, by calling it spiritual food, shows its reference to Christ, as the spiritual drink is explained by the words, "That rock was Christ," which explains the whole. Then is not the cloud and the pillar Christ, who by His uprightness and strength supports our feebleness; who shines by night and not by day, that they who see not may see, and that they who see may be made blind? In the clouds and the Red Sea there is the baptism consecrated by the blood of Christ. The enemies following behind perish, as past sins are put away.[89]

Ephesians 5:25-33.

Husbands, love your wives, just as Christ loved the church and gave himself up for her, in order to make her holy by cleansing her with the washing of water by the word, so as to present the church to himself in splendor, without a spot or wrinkle or anything of the kind – yes, so that she may be holy and without blemish. In the same way, husbands should love their wives as they do their own bodies. He who loves his wife loves himself. For no one ever hates his own body, but he nourishes and tenderly cares for it, just as Christ does for the church, because we are members of his body. "For this reason a man will leave his father and mother and be joined to his wife, and the two will become one flesh." This is a great mystery [*mystērion*], and I am applying it to Christ and the church. Each of you, however, should love his wife as himself, and a wife should respect her husband.

Ephesians 5:31 is a quotation of Genesis 2:24, which Paul interprets as "a great mystery" and explicitly applies to "Christ and the church." Apparently the verse in Genesis carries a deeper meaning. The plain sense

---

[89]Augustine, *Reply to Faustus the Manichaean* 12.29 (NPNF[1] 4:193). See Wilken, *Spirit of Early Christian Thought*, 71-72.

indicates the coming together of a man and a woman in marriage and the love the husband has for his wife. But at the deeper level, Paul tells us that he is applying this to Christ and the church. The details in the text are used "to uncover a truth that is deeper than what is given on the surface of the passage."[90] Once again, the context is not historical; it is theological.

1 Corinthians 10:1-11 and Ephesians 5:25-33. Origen of Alexandria cites both of these passages to counter Celsus's argument that the Bible cannot be interpreted allegorically.[91] These are also the same two passages that Augustine uses at the beginning of *The Literal Meaning of Genesis* to argue for a deeper sense of Scripture. He begins the book by exhorting the reader to consider "what eternal realities are there [in Genesis] suggested, what deeds are recounted, what future events foretold, what actions are commanded or advised."[92] According to Augustine, the Scriptures require special attention because they are divine. Then Augustine asks about the things recounted in Scripture and whether they should be taken as having only a "figurative" meaning or whether they should also be seen as a "faithful account of what actually happened."[93] His answer incorporates the passages in Genesis 2:24, 1 Corinthians 10:11, and Ephesians 5:32:

> No Christian, I mean, will have the nerve to say that they should not be taken in the figurative sense, if he pays attention to what the apostle says: *All these things, however, happened among them in figure* (1 Cor 10:11), and to his commending what is written in Genesis, *And they shall be two in one flesh* (Gn 2:24), *as a great sacrament in Christ and in the Church* (Eph 5:32).[94]

Paul is here used as Augustine's authority in reading Scripture figuratively. Once again, the context is theological rather than historical.

---

[90] Robert L. Wilken, "In Defense of Allegory," *Modern Theology* 14, no. 2 (1998): 200.

[91] Origen, *Against Celsus* 4.49 (ANF 4:520).

[92] Augustine, *The Literal Meaning of Genesis* 1.1.1 (TWSA I/13:168). All quotations of *The Literal Meaning of Genesis* are from Saint Augustine, *The Literal Meaning of Genesis*, trans. Edmund Hill, TWSA I/13 (Hyde Park, NY: New City Press, 2001).

[93] Ibid.

[94] Ibid. Italics in original.

CONCLUDING THOUGHTS ON PAUL
AND THE OLD TESTAMENT

These examples show that Paul had no problem, together with other Jews of the time and later church fathers, recasting Old Testament passages without regard to the author's original meaning of the passage. Paul was the first Christian interpreter of the Old Testament, and he was convinced that "the scriptures speak of, anticipate, typologize, *reveal* Christ and him crucified."[95] Thus, the apostle Paul became the model whom the church fathers sought to emulate in interpretation. His inspired epistles were the warrant for this way of reading the Old Testament Scriptures.

Robert Wilken explains that these three Pauline texts (1 Cor 10; Gal 4; and Eph 5) provided a biblical foundation for a deeper reading of Scripture – a reading that went beyond the plain sense intended by the human author. It is true that the conviction in the early church of the true meaning of what was written in the Scriptures was illuminated by considering history. The history to which this referred, however, was the history that began with and was centered in Christ.

For the church fathers the incarnation was the supreme historical event that gave meaning to all of history, past and present.[96] What we today call the Old and New Testaments were not necessarily thought of as a book.[97] They were, rather, thought of as a twofold event or "covenant," which is not fixed by any written account and unfolds through the ages. The claim by the church fathers that God was the author of Scripture was not made to indicate him as writer but rather as founder and institutor of these two "instruments" of salvation – these two covenants, or economies – which are described in the Scriptures and divide

---

[95]Peter Bouteneff, *Beginnings: Ancient Christian Readings of the Biblical Creation Narratives* (Grand Rapids: Baker Academic), 36. Italics in original.

[96]The following draws from J. Todd Billings, *The Word of God for the People of God: An Entryway to the Theological Interpretation of Scripture* (Grand Rapids: Eerdmans, 2012), 157. See also Henri de Lubac, *Catholicism: Christ and the Common Destiny of Man* (1950; repr., San Francisco: Ignatius Press, 1988), 165-216.

[97]de Lubac, *Catholicism*, 169-83.

between them the history of the world.[98] The crux of this divide, the very meaning of history itself, is Jesus Christ and his incarnation. The incarnation is not merely an idea or event to which Scripture points; it is the key to the meaning of history itself. When the Fathers read Scripture in light of the center of history, Christ, they believed it to be a *historical* reading because the incarnation is the definitive moment in history. The unfolding of God's historical action narrated in the Old Testament thus takes on meaning that would have been inaccessible to the original human writers. What was thought to be history apart from Christ is shown to be what it really is – a "shadow" waiting for fulfillment in Christ. But since Scripture points to Christ as its final end, it is pointing to a mystery that will not be exhausted this side of eternity. Thus, a historical reading of Scripture is necessarily an eschatological one as well, for the church reads Scripture in light of its own union with Christ, which has not yet reached its fulfillment.[99]

To claim that the Old Testament is about Christ and the church means that the story of God's history of Israel does not stand on its own; it is illuminated by Jesus Christ – the key to history – and the church's (eschatologically conditioned) union with him. In the best instances of this union's practice, Abram, Joshua, and Esther do not disappear, only to be replaced with Christ in premodern interpretation. Instead, they are understood in light of history's culmination in the Alpha and Omega, the true human being – Jesus Christ. When the risen Christ "opened the minds" of his companions on the Emmaus Road "to understand the scriptures," he suggested that the "law of Moses, the prophets,

---

[98]See Irenaeus, *Against Heresies* 1.20.1; and 4.32.2 (ANF 1:344-45 and 1:506).

[99]In traditional theological language, eschatology is that branch of theology dealing with last things or end times. In the church fathers, eschatology is not simply the era of earthly time followed by the era of the eternal. Rather, it is a "piercing" of the eternal into the temporal, or perhaps an even better analogy would be the "infection" of the temporal with the eternal. Cf. C. S. Lewis, *Mere Christianity* (San Francisco: HarperSanFrancisco, 2001), book 4, chapter 4, "Good Infection." My thanks to Hanna Lucas, in her Trinity Western University master's thesis ("To Extend the Sight of the Soul: An Analysis of Sacramental Ontology in the Mystagogical Homilies of Theodore of Mopsuestia," 33) for this helpful reference in Lewis and explanation.

and the psalms" have been "fulfilled" in himself (Lk 24:44-45). Robert Wilken sums this up well:

> Its meaning cannot be restricted to what happened in the past. What a text says about past events and persons (things that happened, says Paul) is an integral part of what it means, but the task of interpretation is never exhausted by a historical account. The text belongs to a world that is not defined solely by its historical referent. For St. Paul this is not an enterprise in literary artifice, but a matter of divine revelation. Through Christ it is possible to discern a deeper meaning in the ancient events and to appropriate them "for our instruction."[100]

How, we must ask, do the expectations of the church fathers compare with the expectations of a grammatical-historical approach? Paul, who was clearly not tied to the author's original meaning, allowed later revelation to inform his own understanding of the Old Testament. In doing this, he was used as a model for biblical interpretation by the Fathers. Broad claims that make the church fathers followers of the GH method need to be measured against their dependence and explicit devotion to Paul's "method." Further, patristic exegesis developed in a way that forces us to reconsider simplistic and general conclusions that categorize the Fathers into "literal" and "allegorical." In order to see this, we must understand two important schools of thought in the early church.

ALEXANDRIAN AND ANTIOCHENE EXEGESIS

There is no doubt that a tension existed between the Alexandrian and Antiochene schools of thought. This tension did not exist only in the area of biblical interpretation, but this is where we will direct our focus.[101] The difference between the Antiochene and Alexandrian approaches to Scripture is often expressed as a distinction between typology (Antiochene) and allegory (Alexandrian). The difference between the two was believed to have history at its heart: typology was said to

---

[100]Wilken, "In Defense of Allegory," 201.
[101]The christological controversies of the fourth and fifth centuries were very much associated with the differences in Alexandrian and Antiochene theology.

have "real correspondence to historical events," while allegory "takes no account of history."[102]

The connection of typology to history is reflected quite strongly by some evangelicals. For example, Klein, Hubbard, and Blomberg explain that *typology* is the best term to explain how New Testament writers often used the Old Testament.[103] This insistence is grounded in the presumed understanding of its connection to history. For these authors, R. T. France sets out a clear definition of typology: "the recognition of a correspondence between New and OT events, based on a conviction of the unchanging character of the principles of God's working."[104] In order to distinguish this definition from "abuses of the term typology," the authors emphasize that typology has "correspondence in history" — the correspondence between Old and New Testament *events* are said to be grounded in *history*.[105] The foundation of this connection between typology and history is the authors' belief that "God's ways of acting are consistent through history."[106]

History is such a controlling factor for modern biblical interpretation in Klein, Blomberg, and Hubbard that they actually recommend we *not* use New Testament writers as examples for interpretation today! The Old Testament as a whole is described as having a "forward-looking dimension" to it, of which the original writers were unaware. Later writers of the New Testament then saw divine patterns and made the typological connections. "This view of typology helps us understand what

---

[102]G. W. H. Lampe and K. J. Woollcombe, *Essays on Typology*, Studies in Biblical Theology 22 (London: SCM Press, 1957), 29, 31. For a very helpful account of this historiography, along with a brief critique of the distinction, see Frances Young, "Typology," in *Crossing the Boundaries: Essays in Biblical Interpretation in Honour of Michael D. Goulder*, ed. Stanley E. Porter, Paul Joyce, and David E. Norton (Leiden: Brill, 1994), 29-34.

[103]William W. Klein, Craig L. Blomberg, and Robert L. Hubbard Jr., *Introduction to Biblical Interpretation* (Dallas, TX: Word, 1993), 130.

[104]R. T. France, *The Gospel According to Matthew*, Tyndale New Testament Commentary (Grand Rapids: Eerdmans, 1985), 40.

[105]Klein, Blomberg, and Hubbard, *Biblical Interpretation*, 130. The authors use Klyne Snodgrass ("The Use of the Old Testament in the New," in *New Testament Criticism and Interpretation*, ed. D. A. Black and D. S. Dockery [Grand Rapids: Zondervan, 1991], 416) to make this emphasis.

[106]Klein, Blomberg, and Hubbard, *Biblical Interpretation*, 130.

often occurs when NT writers use the OT in what appear to be strange ways. Certainly they use the OT in ways that we do not recommend to students today!"[107] This is a very curious comment. If the inspired New Testament writers are not recommended as models of proper exegesis for the Christian today, then who indeed should be recommended?

William Tolar, in a handbook on GH biblical interpretation for evangelicals, describes typology as a special kind of biblical interpretation based on the belief that there is a foreshadowing or prefiguring between certain persons, events, and things in the Old Testament with those appearing later in the New Testament.[108] Again, the foundation of typology is God's control of history and his use of these persons or things to point the way to his greater, later revelation in Christ. "Unlike allegory, typology affirms historical reality; but it is figurative in its methodology even while affirming the literalness of its subjects and objects."[109]

For many years the prevailing conclusion was that Antiochenes rejected the Alexandrian approach to biblical interpretation because of their diametrically opposed view of allegory and concomitant emphasis on the literal/historical aspect that was believed to be reflected in typology. Allegory was thought to be poor interpretation because of its callous treatment of history, while typology was thought to be good interpretation because it takes history seriously. In the past half century or so, there has been a reassessment of the traditional antithesis between allegory and typology and its foundation in historical connection. The reassessment has concluded that Alexandria and Antioch represent complementary rather than contradictory or competitive viewpoints.[110]

Reflecting this reassessment, Theodore Stylianopoulos bemoans any sharp distinction between the Alexandrian and Antiochene exegetical

---

[107]Ibid., 131.

[108]William Tolar, "The Grammatical-Historical Method," in *Biblical Hermeneutics: A Comprehensive Introduction to Interpreting Scripture*, 2nd ed., ed. Bruce Corley, Steve W. Lemke, and Grant I. Lovejoy (Nashville: Broadman & Holman, 2002), 28.

[109]Ibid.

[110]Joseph W. Trigg, *Biblical Interpretation*, Message of the Fathers of the Church 9 (Wilmington, DE: Michael Glazier, 1988), 31.

traditions.[111] Rather than seeing these approaches as mutually exclusive, he avers that they are both "fundamentally metaphorical and symbolic." The desire of both approaches in reading Scripture was spiritual edification. By the same token, neither had any desire to abandon the literal sense (as they understood it).

Similarly, Karlfried Froehlich explains that while there is little doubt the Antiochenes did have issues with the excesses of Alexandrian spiritualism, he also warns that it is problematic to make a sharp distinction between Alexandrian and Antiochene exegesis. To claim that only the Alexandrian fathers allegorized while the Antiochene fathers adhered only to the literal meaning of the text is incorrect.[112] In summarizing this line of thinking, Froehlich states:

> In Antioch, the Hellenistic rhetorical tradition, and therefore the rational analysis of biblical language, was stressed more than the philosophical tradition and its analysis of spiritual reality. Moreover, in Alexandria, history was subordinated to a higher meaning; the historical referent of the literal level took second place to the spiritual teaching intended *by* the divine author. In Antioch, the higher *theoria* remained subject to the foundational *historia*, the faithful (or sometimes even fictional) account of events; deeper truth for the guidance of the soul took second place to the scholarly interest in reconstructing human history and understanding the human language of the inspired writers."[113]

Froehlich here identifies some important issues in clarifying patristic exegesis. In particular, we should make note of the important connection he makes between Antioch's rhetorical tradition and Alexandria's philosophical tradition. This general location of influence and adherence is precisely where we can begin to better understand the differences and, especially, similarities that existed between the Alexandrian and Antiochene approaches to biblical interpretation.

---

[111]Theodore G. Stylianopoulos, *The New Testament: An Orthodox Perspective,* vol. 1, *Scripture, Tradition, Hermeneutics* (Brookline, MA: Holy Cross Orthodox Press, 2002), 118.

[112]Karlfried Froehlich, ed., *Biblical Interpretation in the Early Church* (Philadelphia: Fortress Press, 1984), 20.

[113]Ibid., 20-21.

ANCIENT GREEK EDUCATION –
PHILOSOPHY AND RHETORIC

One problem with attributing the tension in the early church to a difference between a literal/historical and an allegorical approach to Scripture is that it assumes the literalism of the Antiochenes is the same as modern historicism. But as Frances Young states, "We can see how this historical emphasis was recognizably culturally specific to the modern world." Antiochenes, Young explains, could not have even imagined

> explicitly locating revelation not in the text of scripture but in the historicity of events behind the text, events to which we only have access by reconstructing them from texts, treating them as documents providing historical data. This is anachronistic, and obscures the proper background of the Antiochene's protest [against allegory].[114]

The proper background for understanding the tension between Alexandria and Antioch is the Greek education system, which was based on the study of literature and practical exercises in speech making. Christianity was inevitably affected by this educational system because of its significant influence on the society and culture into which the early church was born.[115]

The tension that existed between the Antiochenes and the Alexandrians is paralleled in the debate between philosophers and rhetoricians concerning the ideals of education, which is evident in the philosopher Plato and his criticism of the Sophists. Rhetoric, as practiced in the ancient world, was criticized by philosophers because they saw it as having no real purpose except the glorification of the speaker.[116]

---

[114]Frances M. Young, *Biblical Exegesis and the Formation of Christian Culture* (Peabody, MA: Hendrickson Publishers, 2002), 166. I acknowledge my dependence in this section on Young's fine work in this area, particularly 76-81 and 161-85. See also Frances M. Young, "The Rhetorical Schools and Their Influence on Patristic Exegesis," in *The Making of Orthodoxy: Essays in Honour of Henry Chadwick*, ed. Rowan Williams (Cambridge: Cambridge University Press, 1989), 182-99.

[115]Edwin Hatch, *The Influence of Greek Ideas and Usages upon the Christian Church*, 5th ed. (Peabody, MA: Hendrickson, 1995).

[116]Hughes Oliphant Old, *The Reading and Preaching of the Scriptures in the Worship of the Christian Church*, vol. 2, *The Patristic Age* (Grand Rapids: Eerdmans, 1998), 46.

Since orators received much public acclaim, philosophers believed that
one pursued rhetoric in order to be showered in public adoration. We
see this type of accusation in Plato's dialogue *Gorgias*, where Gorgias is
criticized for being a master of technique rather than a visionary for a
cause. Philosophers saw rhetoric as having no moral purpose, which
"had a way of getting lost under the techniques," and as being practiced
by those who only wanted to get ahead in the world.[117]

Conversely, rhetoric had its own negative perspective on philosophy.
It was seen as useless speculation practiced by those who withdrew
from the world. Rhetoricians countered the criticisms of philosophers
by pointing out rhetoric's moral aspects. The aim of rhetoric, they
argued, was actually to prepare students for an active role in civic and
political life. Of these two ideals, rhetoric was definitely more influential. Many people studied rhetoric, while comparatively few had a
philosophical education. Rhetoric was the main agent in the spread of
Greek culture for about eight hundred years.

Frances Young argues persuasively that "philosophy and rhetoric . . .
represent different, though mutually interacting, approaches to texts"
and that we see these approaches to Scripture in the Alexandrians and
Antiochenes.[118] It is often thought that symbolic allegory, as practiced
by philosophers, was the universal way of reading literature in the educational system of the ancient world.[119] But this was not the approach
taken by the grammar and rhetorical schools, which had more influence. The rhetorical approach was intent on deriving ethical models,
useful instruction, and moral principles from the study of literature.[120]

---

[117]Ibid.

[118]Young, *Biblical Exegesis*, 170.

[119]Symbolic allegory is "The tracing of doctrines, or universal truths, or metaphysical and
psychological theories by means of allegory." See Young, "Rhetorical Schools," 183.

[120]Young, *Biblical Exegesis*, 81: "Rhetorical Criticism would always be 'audience-oriented,' looking
for the effect produced. The intention of the author was taken to be the production of that
effect. Most interpretation was anachronistic to the extent that there was little awareness of
the possibility of distinguishing authorial intention and what the interpreter discerned
[D. A. Russell, *Criticism in Antiquity*]. The intent was to effect a response. Literature was expected to be morally edificatory, and the exercise of moral judgement, or krisis, became an

The reaction to Alexandrian symbolic allegory by the Antiochenes was informed by this rhetorical approach rather than a concern for what we would call a GH approach. The Antiochene exegetes had a rhetorical education and certainly would have been influenced by the ideals of that approach to reading literature.[121] In order to understand this influence, it is helpful to briefly trace how students were trained to read literature in the schools of rhetoric, because this inevitably had an effect on how early Christians read Scripture.

Rhetorical training involved teaching the student to attend to the things stated, or the letter of the text (*ta rhēta*), and to understand it according to the word, the letter, or the reading (*pros lexin*).[122] This meant going through several steps:

> So what a teacher does as he reads in class with his pupils the great corpus of classical literature . . . is to analyse a verse into parts of speech, metre [*sic*], etc., to note linguistic usage, especially commenting on acceptable and unacceptable usage and style, to discuss the different meanings which may be given to each word, to expound unusual words, to elucidate figures of speech or ornamental devices, and to 'impress on the minds of the pupils the value of proper arrangement and of graceful treatment of the matter in hand.'[123]

Collectively these steps were called *methodike* (*to methodikon*), which was concerned with the correct reading and organization of sentences and understanding proper senses of words – particularly archaic words in classics like Homer.[124]

---

important aspect of the school tradition. Because of Plato's attack on the poets as morally subversive, echoed indeed by Plutarch who spoke of poetry as a seductive form of deception, much effort was expended in extracting acceptable moral advice from classical texts, and Plutarch's essays *On the Education of Children* and *How the Young Should Study Poetry* demonstrate how text was critically weighed against text, how admonition and instruction were found in tales and myths, and how the inextricable mixture of good and bad in poetry was taken to be true to life, and therefore useful for exercising moral discrimination."

[121]Frances M. Young, *From Nicaea to Chalcedon: A Guide to the Literature and Its Background*, 2nd ed. (Grand Rapids: Baker, 2010), 241-343.

[122]This section on reading literature in rhetorical training is dependent on Young, *Biblical Exegesis*, 76-81; and Young, "Rhetorical Schools," 184-87.

[123]Young, "Rhetorical Schools," 185.

[124]Frances Young, "The Fourth Century Reaction Against Allegory," *Studia Patristica* 30 (1997): 121.

A further aspect of reading in the rhetorical schools was explaining the stories, which entailed clarifying allusions to classical myths, gods, heroes, legends, and histories. Quintilian, a Latin practitioner of Greek rhetoric, offers a caution here to beware of including too much detail for fear of overloading one's mind. This aspect Quintilian calls *historike* (Greek = *to historikon*; *historia*), which for him is technical vocabulary that has to do with knowledge acquired by investigation and enquiry and does not necessarily imply any claim to historical factuality in our modern sense. Young explains that

> [*Historikon*] did not attempt to distinguish historical background as such from other bits of erudite information which might throw light on the text – it embraced astronomy, geography, natural history, music, any-thing and everything, including myths, and legends, that might satisfy the curiosity of the enquirer ([*historein*] is, after all, "to enquire").[125]

This should caution against reading, in *historikon*, something akin to grammatical-historical analysis.[126] Ancient literary criticism really had no true historical sense.[127]

We should make one qualification regarding the comments about history. In Herodotus there is an example of the ancient literary genre of *historia* employed in a way not totally dissimilar to what we mean by history. Known as the "father of history," Herodotus compiled the results of his inquiries into a narrative, which had many subnarratives sometimes told in different versions, with no real attempt to distinguish between them. These also included local "myths," and (from our perspective) a significant amount of his information would not be regarded as historical but rather as geographical or cultural. Successors of Herodotus, like Thucydides and Polybius, claimed to restrict themselves to only "the facts," but it was very common to invent speeches of main

---

[125]Ibid.

[126]Young adds, "I submit that too many discussions of patristic exegesis have jumped to conclusions about historical interest where such terminology is used, for it does not necessarily imply what we mean by historical." See *Biblical Exegesis*, 79.

[127]D. A. Russell, *Criticism in Antiquity* (London: Duckworth, 1981), chaps. 8 and 11.

characters when no information was available. Even Thucydides admitted to having composed such speeches.[128]

The literary genre called *historia* did mean deeds (*pragmata*) or things that happened (*res gestae*), and stories labeled as *historia* were expected to be "true." But this was not the distinctive thing about them.[129] There was far more importance placed on the moral significance of things that happened. Further, as Young notes, this presentation needed to be composed creatively, with effective style:

> It is no accident that our word "story" is derived from *historia*; for the narrative was meant to be a good, improving story (*ktēma es ai* – a "possession for ever," according to Thucydides [1.22 LCL, vol. 1, p. 40]). *To historikon* involved the investigation of the "story" presented in the text being studied.[130]

This did not mean that no critical questions were asked of the narrative.[131] In fact, there existed a distinction between three types of narrative: true history, or an accurate account of real events; fiction, or what could

---

[128]Thucydides, *History*, 4 vols., trans. Charles Foster Smith, LCL (Cambridge, MA: Harvard University Press, 1919–1923), 1.22.

[129]Young states: "The distinctive thing about historical writing was not 'single-minded pursuit of facts.'" *Biblical Exegesis*, 166-67.

[130]Ibid., 80.

[131]Young, "Fourth Century Reaction Against Allegory," 123-24 states: "But clearly the larger issue of how history was understood in the ancient world cannot be passed over. The fact that Lucian could write a treatise objecting to the rhetorical excesses of writers claiming to produce history (Lucian, *How to Write History*, ed. and Eng. tr. by K. Kilburn, Loeb Classical Library, vol. vi of Lucian's works, pp. 2-72) confirms both the point that ancient readers of this genre expected to be told 'true stories' in an 'objective' way (M. J. Wheeldon, '"True Stories": the reception of historiography in antiquity,' in *History as Text*, ed. Averil Cameron [London, 1989], pp. 36-63) – though clearly they often got questionable ones; indeed, novelistic romances parodied the conventions of 'history' (Niklas Holzberg, *The Ancient Novel*, An Introduction [Eng. tr. by Christine Jackson-Holzberg, London, 1995]) – and also the point that plain factuality was not regarded as the aim either. Even Lucian suggests that the historian, 'if a myth comes along,' should tell it but not believe entirely: his advice is 'make it known for the audience to make of it what they will.' History was a descriptive narrative intended to improve as much as inform. The genre embraced not just past events, but all kinds of other information – geographical, cultural, technical, strategic, you name it; and it was supposed to be useful, to explore moral issues, and the interplay of fate and fortune in the affairs of men. The Antiochenes could not have had the anxiety about historicity that has bothered modern scholars, with their detective model of historical research. Nevertheless, there was ancient discussion about the plausibility of narratives."

have happened but did not; and myth, what could not have happened, or a false account. There are several things that may render an account as myth, but the methods of criticism were based in comparison and plain logic. Incredulity in a narrative may be "because of the character to which the action is attributed, the nature of its action, its time, its mode of performance, or the motive or reason adduced. There may be omissions or contradictions in the narrative which suggest the author is not to be trusted."[132] This often caused the critic to resort to allegorization.

At this point studies come to be geared more toward the practice of composition. Texts are assessed and distinguished between the subject matter (*ho pragmatikos topos* or *res*) and the style and vocabulary (*ho letikos* or *onomata verba*). Successful communication of object lessons, the author's skill in organizing and manipulating the subject matter, and his choice of presentation style were very important. It was, however, the interest in style that predominated.

Good style led to clear communication, so attention was paid to the schema of a speech or writing.[133] A well-constructed and deliberate schema was developed in order to prove one's case or argument. This included beginning with an *exordium*, or introduction. After this the main material was presented as *narratio*, or the statement of the proposition (*peroratio*) that the speaker or orator sought to maintain. This was followed by arguments in favor of the *peroratio*, which are called *confirmatio*. Arguments against the *peroratio* were presented after the *confirmatio*. All this was concluded with another *peroratio*, or conclusion, in support of the original *peroratio*.

The exegesis practiced in the schools was very much concerned with the outer dress of diction and style. Language was understood to be the clothing for the thoughts, which required enunciation and distinction. It was assumed that the author had a subject matter to cover and a thesis to assert, so exegesis included discerning this thesis. However, the

---

[132]Young, *Biblical Exegesis*, 80.
[133]This paragraph draws on Old, *Reading and Preaching of the Scriptures*, 2:49.

criticism was audience centered. From our perspective, there was a certain anachronism about it. Distinguishing between what the text meant (author's original meaning) and what the text means (application) was not a concern.[134] Interest was more in the effect the text produced, which is why we cannot draw a straight line from this approach to grammatical-historical or even historical-critical methods of interpretation.[135] Since the intent of criticism was to effect a response, ancient exegetes expected literature to be morally uplifting. This entailed the exercise of moral judgment (*krisis*), which included literary, or rhetorical, evaluation. Questions of authenticity, dating, and the like were raised here, but it was much less critical in our sense, and the moral search for virtue was predominant.

Already by about the sixth century BCE the two works attributed to the Greek poet Homer, the *Iliad* and the *Odyssey*, had formed the basis for Greek culture and education.[136] But the stories contained in these works contained portraits of the gods that were considered scandalous by later writers. Plato is well known for his critique of the myths of the poets and actually banned them from his republic. His reasoning for this is because "they made up untrue stories which they used to tell people—and still do tell them."[137] He argued that stories like the one in Hesiod's *Theogony* about Kronos have a detrimental effect on the young, who lack the critical faculties to discern them.[138] The general

---

[134]Russell, *Criticism in Antiquity*.

[135]Young, "Rhetorical Schools," 186.

[136]Mark Sheridan, *Language for God in the Patristic Tradition: Wrestling with Biblical Anthropomorphism* (Downers Grove, IL: IVP Academic, 2015), 45. What follows draws on pp. 45-55 of Sheridan.

[137]Plato, *The Republic* 377d. All quotations from *The Republic* are from Plato, *The Republic*, trans. G. R. F. Ferrari and Tom Griffith (Cambridge: Cambridge University Press, 2000).

[138]The story of Kronos is told in Hesiod, *Theogony* 126, 172, 453. Kronos was the Titan god of time who castrated his father and then, out of fear that he would be overthrown by his own son, swallowed each of his children as soon as they were born to him by Rhea. She, however, was able to save one son, Zeus, by giving Kronos a stone wrapped in swaddling clothes instead of the infant. When Zeus grew up, he forced Kronos to disgorge his swallowed children, and led the Olympians in a ten-year war against the Titans, eventually defeating them and driving them into the pit of Tartaros. Referring to this story, Plato states in *The Republic*: "As for what Kronos did, and what his son did to him, even if they were true I wouldn't think that in the normal course of events these stories should be told to those who are young and uncritical. The best thing would be to say nothing about them at all" (378a).

attitude of Plato toward these myths is expressed in the person of Socrates:

> Nor, in general, any of the stories – which are not true anyway – about gods making war on gods, plotting against them, or fighting with them. Not if we want the people who are going to protect our city to regard it as a crime to fall out with one another without a very good reason. The last thing they need is to have stories told them, and pictures made for them, of battles between giants, and all the many and varied enmities of gods and heroes towards their kinsmen and families.[139]

Plato's main problem was that these myths give a false view of the nature of divinity – gods simply would not behave in the way described in the myths.[140] Against the mythical discourse of the poets, Plato thus argued for philosophical discourse.

The significance of critiques like Plato's against Homer is seen in the fact that even supporters of Homer were in basic agreement with them. It is noteworthy, however, that Homer was not simply abandoned. The likely reason is because Homer was so deeply ingrained in Greek culture, as is seen in Heraclitus, one of his first-century defenders:[141]

> From the very first age of life, the foolishness of infants just beginning to learn is nurtured on the teaching given in [Homer's] school. One might almost say that his poems are our baby clothes, and we nourish our minds by draughts of his milk. He stands at our side as we each grow up and shares our youth as we gradually come to manhood; when we are mature, his presence within us is at its prime; and even in old age, we never weary of him. When we stop, we thirst to begin him again. In a word, the only end of Homer for human beings is the end of life.[142]

But even Heraclitus is aware of the problems in Homer: "If he meant nothing allegorically, he was impious through and through, and sacrilegious fables, loaded with blasphemous folly, run riot through both

---

[139]Plato, *The Republic* 378c.
[140]Sheridan, *Language for God*, 50.
[141]Ibid. The quote from Heraclitus below is also found in Sheridan.
[142]Donald A. Russell and David Konstan, *Heraclitus: Homeric Problems*, Writings from the Greco-Roman World (Atlanta: Society of Biblical Literature, 2005), 3.

epics." Thus, the solution offered by Heraclitus, and others who felt Plato's critique, was to interpret Homer allegorically.[143]

Plutarch (ca. 46–120 CE) also recognized Homer's importance in Greek education[144] and therefore advises certain ways of reading to find "the seeds of discourse and action"[145] in Homer that produce "persuasive or educative effect through those parts that are useful in the pursuit of virtue."[146] According to Young, Plutarch believed that poetry was a seductive form of deception that required a kind of "moral 'pruning' which lays bare the profitable things that are hidden under the prolific foliage of poetic diction and clustering tales."[147] Thus, Homer was read allegorically as containing the whole of philosophy: "This is the road to understanding the nature of reality and of divine and human matters, to distinguishing on the ethical plane those things which are good and those which are bad, and to learning any rule of reasoning appropriate for reaching the truth."[148]

Plutarch's "moral criticism" was taken and used by the rhetorical schools.[149] It was important that the student, under the tutelage of the teacher (*grammaticus*), become comfortable with reading poetry in order to bring it out of the realm of myth. This meant expending much effort in extracting acceptable moral advice from classic texts. Thus, texts were weighed against each other to investigate how admonition and instruction were found in these myths. Students were taught that the "mixture of good and bad in poetry" was imitative of life and thus important and "useful for exercising moral discrimination."[150]

---

[143]"As the originator of all wisdom, Homer has, by using allegory, passed down to his successors the power of drawing from him, piece by piece all the philosophy he was the first to discover." Russell and Konstan, *Heraclitus*, 63, cited in Sheridan, *Language for God*, 51.

[144]"It is appropriate that Homer, who in time was among the first of poets and in power was the very first, is the first we read. In doing so we reap a great harvest in terms of diction, understanding, and experience of the world." Plutarch, *Essay on the Life and Poetry of Homer*, ed. John J. Keaney and Robert Lamberton (Atlanta: Scholars Press, 1996), 67n1, cited in Sheridan, *Language for God*, 53.

[145]Plutarch, *Essay on the Life and Poetry of Homer*, 71n6.

[146]Ibid., 71.

[147]Young, "Rhetorical Schools," 187.

[148]Plutarch, *Essay on the Life and Poetry of Homer*, 157n92, as cited in Sheridan, *Language for God*, 54.

[149]Young, "Rhetorical Schools," 187.

[150]Young, *Biblical Exegesis*, 81.

It should be evident that this approach to ancient literature could be very conducive to the development of allegory. It is clear, however, that ancient rhetorical training was not characterized by what is called *symbolic allegory*.[151] The kind of allegory that sought cosmological or mystical meaning (philosophical) was rejected in favor of moral lessons (rhetorical). This has prompted Frances Young to indicate an important distinction in the ancient allegorical reading of texts: "Perhaps we should conclude there was allegory and then there was allegory—reading texts symbolically and mystically was philosophical whereas reading texts to tease out a moral was rhetorical."[152]

PHILOSOPHY AND RHETORIC MEET IN ALEXANDRIA AND ANTIOCH

In returning to the reason for this excursion through ancient Greek education, we can see its relevance to the Antiochene and Alexandrian issue at hand. When it came to reading texts, philosophy and rhetoric were two overlapping yet distinct traditions. There were linguistic, textual, and etymological interests held in common, and philosophy built on the techniques of a school's exegesis. Differences lay, however, in their attitudes to what a text's fundamental meaning and reference was, as well as how this meaning and reference could be discerned. Philosophy looked for abstract doctrines or virtues through verbal allegory, while rhetoric looked for solid ethical examples in reading narrative. Further, in writing and presenting, rhetoric worked toward modeling stylistic and ethical excellence.

The Alexandrian Origenist tradition adopted the allegorical techniques of the philosophers, while the Antiochenes reacted in protest, in the tradition of rhetoric, against the Alexandrian way of handling texts.[153] This, once again, should reveal an incongruity with the typical

---

[151]Young states that this was practiced by the Stoics ("Rhetorical Schools," 188).
[152]Ibid.
[153]Ibid.

appropriations of the Alexandrian/Antiochene disagreement. Antiochene exegesis is anchored in rhetoric, and this makes any attempt to characterize it as the precursor to the modern critical approach problematic. To speak of any sort of grammatical-historical exegesis in antiquity is actually anachronistic, and equating the *historia* employed by Antiochenes with our modern understanding of history is equally so.

It is clear that the great Antiochene exegetes all received an education in rhetoric.[154] When one turns to the actual exegesis practiced by these Antiochenes, numerous examples confirm that rhetorical training had significant influence on how they read Scripture.[155] Space does not permit an extended examination of these Antiochene examples of exegesis, but it is quite different from modern grammatical-historical approaches. Foundationally, we see that it was not a historical concern that produced the Antiochene reaction to allegory. Antiochenes were very concerned to take the text seriously, which they believed a word-for-word spiritual allegory failed to do. This is one of the reasons that many readers of Antiochene exegesis comment on the "dull and pedestrian character of Antiochene commentaries."[156] The commentaries remain concerned predominantly with paraphrase and the "nitty gritty" of the text and rarely rise above that.

ANTIOCHENE EXEGESIS – DIODORE OF TARSUS

It may be helpful here to summarize some aspects of one famous Antiochene exegete because it will help us better understand both the Antiochene attitude to history and the concerns they had with allegory. In the prologue to his *Commentary on the Psalms* we see Diodore of Tarsus exegeting the text with concern for the *methodikon* and

---

[154]Examples can be seen in Young, *Biblical Exegesis*, 171; and "Rhetorical Schools," 189.
[155]Young gives many examples in *Biblical Exegesis*, 171-75; and "Rhetorical Schools," 190-93.
[156]Young, "Rhetorical Schools," 190.

historikon of rhetoric.[157] His exposition of the Psalms, he states, is "an explanation of their plain text."[158] The reader of the Psalms needs to grasp "the logical coherence of the words." Anchoring his exposition in the plain text is important for Diodore because it will keep the worshiper's mind from being "carried away by the words."[159] A little later in the prologue, Diodore emphasizes this anchoring in the plain text once again: "We will not shrink from the truth but will expound it according to the historical substance (*historia*) and the plain literal sense (*lexis*)."[160] But the *historia* and *lexis* do not cause Diodore to disparage the higher (spiritual) sense of Scripture, the *theōria*. History, Diodore continues, is not opposed to *theōria* but rather is its foundation – it is the basis of the higher senses of Scripture. If *theōria* loses the connection to the *historia* and *lexis*, it ceases being *theōria* and becomes allegory. It is important here not to read our understanding of history into the ancient concern for *historia*. As we have been reminded above, ancient literary criticism really had no true historical sense in the way we understand it.[161]

MIMĒSIS

The previous section exposed an important difference between Antiochene and Alexandrian exegesis. Antiochenes argue, as Diodore above, that the narrative, the story, or the text (not necessarily the historical events to which the narrative refers) is essential in any move toward a higher reading. One of the more significant issues they had with allegory was that it ceases being story.[162] Frances Young argues that the difference between Antiochene and Alexandrian approaches to Scripture

---

[157]Diodore of Tarsus, "Commentary on the Psalms, Prologue," in *Biblical Interpretation in the Early Church*, Sources of Early Christian Thought, ed. Karlfried Froehlich (Philadelphia: Fortress Press, 1984), 82-86.

[158]Ibid., 83.

[159]Ibid.

[160]Ibid., 85.

[161]Russell, *Criticism in Antiquity*, chaps. 8 and 11.

[162]Young, *Biblical Exegesis*, 161: "Allegory ceases to be story and becomes propositional; typology, on the other hand, retains the narrative and the sequence."

is in the way that perceived deeper meanings were taken to relate to the surface of the text.[163] She thus proposes a distinction between "ikonic *mimēsis*" (Antiochene) and "symbolic *mimēsis*" (Alexandrian).[164]

This distinction points to the Antiochene expectation of finding a real connection between what the text stated and the higher sense. Here *mimēsis* should be understood in the sense of representation.[165] Something could be represented through genuine likeness, "ikon" or image, or an analogy, or it could be represented by a symbol, something unlike that which stands for the reality.

> The "ikon" will resemble the person or event which it represents, but symbols are not representations in that sense; symbols are "tokens" or "signs" whose analogous relationship with what is symbolized is less clear. Indeed, whereas a text offers clues to its "ikonic" intention, the clues to "symbolic" language lie in the impossibilities which cannot be taken at face value.[166]

This is precisely what Diodore is getting at in his insistence that the higher sense, the *theōria*, not lose its connection to the *historia* and the *lexis*. If it does, then it ceases being *theōria* and becomes allegory – it ceases being story and moves into the realm of proposition. That is, it ceases being ikonic and becomes symbolic.

The distinction centers on a different perception of how Scripture related to its referent.[167] Both expected a deeper sense, and neither was concerned with the reference to the events behind the text or the human author's intended meaning. The Antiochene (ikonic) approach expects a mirroring or imaging of the deeper meaning in the text as a whole, while the Alexandrian (symbolic) approach was seen by the Antiochenes as destroying the story, or coherence, of the text because it

---

[163]Young, "Fourth Century Reaction Against Allegory," 123.
[164]Young, *Biblical Exegesis*, 161-62.
[165]Ibid., 210-11.
[166]Ibid., 210.
[167]Ibid., 162.

involved using words as symbols or tokens.[168] "What is different is the [Antiochene] assumption that the narrative provides a kind of 'mirror' which images the true understanding, rather than the words of the text providing a code to be cracked."[169]

The issue is much deeper than the simple conclusion that the Antiochene insistence on typology was the result of its historical anchoring in events, while Alexandrians preferred allegory because of their disdain for history.[170] The fact is that all early Christian reading of Scripture is, in some sense, figural. The nature of figural language, as described by Erich Auerbach, can help us here:

> Figural interpretation establishes a connection between two events or persons in such a way that the first signifies not only itself but also the second, while the second involves or fulfills the first. . . . A connection is established between two events which are linked neither temporally nor causally — a connection which it is impossible to establish by reason in the horizontal dimension. . . . It can be established only if both occurrences are vertically linked to Divine Providence, which alone is able to devise such a plan of history and supply the key to its understanding.[171]

Here Auerbach makes reference to horizontal and vertical linking, which requires some explanation.[172] He writes of Christian figural reading being dependent on access to a hermeneutical key that lay

---

[168]In "Fourth Century Reaction Against Allegory," Young puts it this way: "[T]he difference between Antiochene and Alexandrian exegesis lies in the way perceived deeper meanings were taken to relate to the surface of the text. The difference may be characterized as that between an 'ikonic' and a 'symbolic' relationship. An 'ikon' represents and images the underlying reality, a 'symbol' is a token, with no necessary likeness. Allegory took words as discrete tokens, and by de-coding the text found a spiritual meaning which bore no relation to the construction of the wording or narrative. Antiochene exegesis embraced typology and prophecy, morals and dogma, but only by allowing that the sequence of the text mirrored or imaged the realities discerned by [*theōria*]. The whole [*schēma*] is important" (123).

[169]Young, *Biblical Exegesis*, 162-63.

[170]John J. O'Keefe, "'A Letter That Killeth': Toward a Reassessment of Antiochene Exegesis, or Diodore, Theodore, and Theodoret on the Psalms," *Journal of Early Christian Studies* 8, no. 1 (2000): 94.

[171]Eric Auerbach, *Mimesis: The Representation of Reality in Western Literature*, trans. Willard R. Trask (Princeton, NJ: Princeton University Press, 1953), 73, cited in O'Keefe, "'Letter That Killeth,'" 94-95.

[172]This commentary on Auerbach draws from O'Keefe, "'Letter That Killeth,'" 95.

outside the normal connection between events. Figurative reading sought to allow that connection to be made and to enable the reader to be more than merely a reader. It sought to make it possible for the reader to enter the narrative world of the text *as a participant*. A reading of Scripture that seeks to anchor texts in their historical events maintains a certain gap between reader and text. But figural readings "render the lives of ordinary people more significant and more real because, through figuration, the biblical text can speak directly and concretely."[173]

For example, one can speak about the connection between the Israelites crossing the Red Sea after their exodus from Egypt and the Christian crossing from sin to new life through baptism as a vertical connection.[174] There is no necessary connection between the two on a historical plane in a sequence, so it is not a horizontal connection. The connection is vertical through God's plan for the salvation of humanity. While a horizontal connection can be established, and the events can be morally exemplary, the events still remain distant and not given to imitation. This vertical connection is fundamentally different from the common sequence of events in classical historiography. But the vertical sequencing allows the event to escape the realm of history and thus allows Christians "to see the reality of their lives as somehow enhanced by the story of Exodus. Exodus is more than a heroic tale of God's past deeds and ancient courage; it is the tale of my deliverance from sin and death."[175]

Recall here John MacArthur's condemnation of his very first sermon, based on the text "And the angel rolled the stone away," in which he preached about the stone of doubt, the stone of fear, and the stone of anger.[176] MacArthur condemned the sermon because he believes that in doing this he had betrayed the historical referent – "That is not what that verse is talking about; it's talking about a real stone."[177] The emphasis on

---

[173]Ibid.

[174]Example from ibid., 95.

[175]Ibid., 95.

[176]See chapter three under the heading "The Inappropriateness of Allegory." See also MacArthur, *How to Study the Bible*, loc. 1163-64.

[177]Ibid.

the "real stone" has kept MacArthur anchored in the past, the horizontal plane, and thus maintains the distance and gaps that he emphasizes throughout his entire book. But a vertical reading seeks to eradicate those gaps by inviting the interpreter to be a participant. I dare say that the Alexandrians would have actually commended MacArthur's reading of the text in his first sermon. Surely this example is not making the Bible into a fairy-tale book from which we get "all kinds of crazy interpretations."

Because of its grounding in classical rhetoric, the Antiochene approach resisted this kind of vertical connection. The rhetorical grounding did not predispose them to a wholesale rejection of figural readings of Scripture. Rather, they were unwilling to use it vertically. They often employed figural language horizontally in order to fit a text into the construct of the chronicle, but they rejected any vertical connection.[178]

ORIGEN AND EUSTATHIUS: THE CLASSIC EXAMPLE

The classic example of the difficulty of clearly dividing the Alexandrians and the Antiochenes into strictly allegorical versus literal/historical interpreters is Eustathius of Antioch's (ca. 270–ca. 337) extended critique of Origen of Alexandria's (ca. 185–ca. 254) interpretation of 1 Samuel 28:5-18. Both Origen's original interpretation and Eustathius's critique have attracted many who see in them the traditional distinction between Antiochene literalism and Alexandrian allegory.[179] Patristic scholars, however, have good reason to question this conclusion.

---

[178]For example, in their commentaries on the Psalms, Diodore and Theodore view the Psalms as essentially a poetic gloss on the history of Israel at the time of David. Thus, the reference point for the Psalms was the book of Kings because it contained the *historia* of David's life. This, in itself, was not necessarily uniquely Antiochene in practice. The characteristically Antiochene thing about it is that the narrative of Kings completely controls the meaning of the Psalms. Thus, for Diodore and Theodore, the Psalms are "literally" about the events that are more clearly narrated in Kings. Thus, the *skopos* of Psalms was limited to the book of Kings and could not be extended to speak of Christ, the church, and the entire Christian life. See O'Keefe, "'Letter That Killeth,'" 93.

[179]Joseph W. Trigg, "Eustathius of Antioch's Attack on Origen: What Is at Issue in an Ancient Controversy?," *Journal of Religion* 75 (1995): 220. See, for example, Johannes Quasten, *Patrology*, vol. 3, *The Golden Age of Greek Patristic Literature* (Westminster, MD: Christian Classics, 1986), 303; R. V. Sellers, *Eustathius of Antioch and His Place in the History of Christian Doctrine* (Cambridge: Cambridge University Press, 1928), 81; D. S. Wallace-Hadrill, *Christian Antioch: A Study of Early Christian Thought in the East* (Cambridge: Cambridge University Press, 1982).

Although the church historian Sozomon describes Eustathius as having written many works, today we possess only his *On the Belly-Myther, Against Origen*, the Antiochene school's first response to Origen's exegesis.[180] It was written at the request of a now unknown Eutropius, whom Eustathius describes as a "most distinguished and holy preacher of orthodoxy."[181] Eutropius had requested Eustathius's position with regard to the belly-myther from 1 Samuel 28 because of a dissatisfaction with what Origen had published on it many years earlier.[182] Eustathius indicates that he is aware of others who find fault with what Origen has written "so off-handedly," and some were even led astray by it.[183] Eustathius honors this request and sends Eutropius Origen's homily along with his own "explanation of the text" so as "through each to make evident the plain sense."[184] Later he invites his readers along as he examines "the very letter of the narrative."[185] The interesting thing here is that Origen describes his own approach to the passage in similar terms as one that is "in accordance with the word."[186] The context in which the controversy between Origen and Eustathius is often framed raises the question about literal interpretation. Each believed that their individual approach to 1 Samuel 28 was literal according to the letter, word, or plain sense. But, as stated above, Eustathius's treatise has been described as accusing Origen of being "against the literal" and rejecting his "entire allegorical exegesis."

In the treatise itself Eustathius accuses Origen of piling up "instruments of idolatry and necromancy" into the church.[187] Origen is

---

[180]Eustathius of Antioch, *On the Belly-Myther, Against Origen*, in *The "Belly-Myther" of Endor: Interpretations of 1 Kingdoms 28 in the Early Church*, ed. Rowan A. Greer and Margaret M. Mitchell, Writings from the Greco-Roman World (Atlanta: Society of Biblical Literature, 2007), 62-157.

[181]Eustathius, *On the Belly-Myther* 1.1 (Greer and Mitchell, 63).

[182]"Belly-myther" is the translation for the Greek word *engastrimython*. Both Eustathius and Origen used the Greek translation of the Old Testament Scriptures called the Septuagint (LXX). In the NRSV, it is translated "medium." See Origen of Alexandria, *Homily 5 on 1 Kingdoms*, in Greer and Mitchell, *"Belly-Myther" of Endor*, 32-61.

[183]Eustathius, *On the Belly-Myther* 1.2-3 (Greer and Mitchell, 63).

[184]Ibid., 1.4 (Greer and Mitchell, 63). The Greek here is *grammatos hypagoria*, which refers to an explanation of the letter.

[185]Ibid., 2.1 (Greer and Mitchell, 65), *to tēs historias gramma*.

[186]I.e., "literally" (*kata ton logon*). Origen, *Homily 5 on 1 Kingdoms* 2.3 (Greer and Mitchell, 35).

[187]Eustathius, *On the Belly-Myther* 3.4 (Greer and Mitchell, 69).

variously accused of repeating himself "like an old crone," "stooping to deceptive artifice," and coming to his conclusions through a "drunken folly."[188] In a particularly revealing section, Eustathius criticizes Origen for "allegorizing all the scriptures."[189] In that section Eustathius voices his displeasure over Origen's allegorization of several passages of Scripture, but this is not the basis on which he expresses his specific criticism of Origen's interpretation of 1 Samuel 28.[190] Indeed, there is no explicit accusation against Origen's exegesis of this passage as allegorical. In fact, Eustathius is puzzled as to why Origen does *not* allegorize: "Accordingly, though he took in hand to allegorize all the scriptures, he does not blush to understand this passage alone according to the letter [*grammatos ekdexesthai*], declaring his interpretation hypocritically, even though he does not pay attention to the body of scripture right-mindedly."[191]

Eustathius does not attack Origen's allegorical interpretation of 1 Samuel 28 but rather his *literal* interpretation. Understanding why this is so can help us put the issues between Alexandria and Antioch in their proper context and move beyond the simplistic and ahistorical division of the controversy as one of literal/historical (in our modern historicist sense) versus allegorical.

The narrative at issue is the culmination of the decline of Saul. In obedience to God, Saul had already sought to expel mediums and wizards from Israel. Yet now, as the Philistine army is gathered around him, he seeks guidance from the very people he had sought to expel:

> When Saul saw the army of the Philistines, he was afraid, and his heart trembled greatly. When Saul inquired of the LORD, the LORD did not answer him, not by dreams, or by Urim, or by prophets. Then Saul said

---

[188]Ibid., 17.1-2 (Greer and Mitchell, 117); and ibid., 20.2 (Greer and Mitchell, 125).

[189]Ibid., 21.1 (Greer and Mitchell, 127).

[190]Ibid., 21.1-12 (Greer and Mitchell, 125-31).

[191]Ibid., 21.1 (Greer and Mitchell, 125). The accusation that Origen "does not pay attention to the body of scripture right-mindedly" is a rebuttal to Origen's exhortation that "it is necessary to be right-minded in hearing the scriptures." See Origen, *Homily in 1 Kingdoms 28* 4.1 (Greer and Mitchell, 39).

to his servants, "Seek out for me a woman who is a medium, so that I may go to her and inquire of her." His servants said to him, "There is a medium at Endor." So Saul disguised himself and put on other clothes and went there, he and two men with him. They came to the woman by night. And he said, "Consult a spirit for me, and bring up for me the one whom I name to you." The woman said to him, "Surely you know what Saul has done, how he has cut off the mediums and the wizards from the land. Why then are you laying a snare for my life to bring about my death?" But Saul swore to her by the LORD, "As the LORD lives, no punishment shall come upon you for this thing." Then the woman said, "Whom shall I bring up for you?" He answered, "Bring up Samuel for me." When the woman saw Samuel, she cried out with a loud voice; and the woman said to Saul, "Why have you deceived me? You are Saul!" The king said to her, "Have no fear; what do you see?" The woman said to Saul, "I see a divine being coming up out of the ground." He said to her, "What is his appearance?" She said, "An old man is coming up; he is wrapped in a robe." So Saul knew that it was Samuel, and he bowed with his face to the ground, and did obeisance. Then Samuel said to Saul, "Why have you disturbed me by bringing me up?" Saul answered, "I am in great distress, for the Philistines are warring against me, and God has turned away from me and answers me no more, either by prophets or by dreams; so I have summoned you to tell me what I should do." Samuel said, "Why then do you ask me, since the LORD has turned from you and become your enemy? The LORD has done to you just as he spoke by me; for the LORD has torn the kingdom out of your hand, and given it to your neighbor, David. Because you did not obey the voice of the LORD, and did not carry out his fierce wrath against Amalek, therefore the LORD has done this thing to you today.[192]

While the passage does come across as a straightforward historical narrative, it raised many problems for the church fathers.[193] Patricia Cox summarizes it well: "A narrative that has a witch, the ghost of a dead prophet, and a frightened king who breaks his own law is certainly

---

[192] 1 Sam 28:5-18.

[193] Trigg, "Eustathius of Antioch's Attack," 222. The passage also raised significant problems for Jews. For a fine overview of how both Jews and Christians dealt with the passage and problems it raised, see K. A. D. Smelik, "The Witch of Endor: 1 Samuel 28 in Rabbinic and Christian Exegesis Till 800 A. D.," *Vigiliae Christianae* 33 (1977): 160-79.

an unusual, if not bizarre, history. Its oddness did not escape Patristic exegetes, who for the most part found themselves affronted by the scene so realistically and economically depicted there."[194] Similarly, K. A. D. Smelik states, "The main objection to a literal interpretation . . . is that one has to presume a wizard to be able to raise a most holy prophet from the dead which is impossible, since the ministers of Satan can never be more powerful than the ministers of God."[195]

If the text were read as literal history it would apparently affirm the following:[196]

1. The prophet Samuel was in Hades.

2. A medium was an appropriate intermediary between the human and divine worlds.

3. Magic works, even on the holy people of God.

Many church fathers recognized the significance of these issues, and their subsequent interpretations fall into one of three basic views:[197]

1. Samuel was resuscitated by the medium.[198]

2. Either Samuel or a demon in his shape appeared at God's command.[199]

3. A demon deceived Saul and gave him a forged prophecy.[200]

---

[194]Patricia Cox, "Origen and the Witch of Endor: Toward an Iconoclastic Typology," *Anglican Theological Review* 66, no. 2 (1984): 139.

[195]Smelik, "Witch of Endor," 166.

[196]Cox, "Origen and the Witch of Endor," 139. Trigg ("Eustathius of Antioch's Attack," 222) puts it this way: "Was a departed prophet subject, against his will to a medium [*engastrimythos* in the LXX] and her presumed demonic accomplices? Could a righteous prophet be expecting a wicked king to join him shortly in hell? Could necromancy provide accurate knowledge of the future, and if so might it not be permissible to resort to it?"

[197]Smelik, "Witch of Endor," 164-65.

[198]This was the view of Justin Martyr, Origen, Zeno of Verona, Ambrose, Augustine, Sulpicius Severus, Dracontius, and Anastasius of Sinaita.

[199]This was the view of John Chrysostom, Theodoret of Cyrrhus, Pseudo-Justin, Theodore bar Koni, and Isho'dad of Merv.

[200]This was the view of Tertullian, Pseudo-Hippolytus, "Pionius," Eustathius of Antioch, Ephraem, Gregory of Nyssa, Evagrius Ponticus, Pseudo-Basil, Jerome, Philastrius, Ambrosiaster, and Pseudo-Augustine.

Eustathius, who held the third view, and Origen, who held the first view, devoted the most attention to the passage.

The actual details of the exegesis of Eustathius and Origen are less of a concern for us than the question of why Origen the Alexandrian refuses to allegorize the text, while Eustathius the Antiochene refuses to take the text at face value. Eustathius cannot accuse Origen of allegorizing the text, yet he still finds fault with Origen's interpretation of it. Ultimately, the difference between the two is not allegory but rather a different mentality, which is seen in a fundamentally different understanding of the role of Scripture.[201]

ORIGEN'S APPROACH TO THE PASSAGE

The first thing Origen does is clarify the importance of the passage. Some narratives (*historiai*), he states, are irrelevant for us, while others "are necessary for our hope," and this specific passage "touches all people."[202] The reason it touches all people is that it relates to the fate of Christians after death – this is his ultimate concern in the passage.[203] Origen describes his approach as "narrative" – that is, *historiai* – in order to differentiate it from what he calls the "elevated sense" (*anagōgēs*).[204] The narrative of Lot and his daughters (Gen 19:30-38), for example, has benefit according to the elevated sense, which is granted by God. Origen asks explicitly, "What is the benefit for me from the narrative about Lot and his daughters? Likewise, what is the benefit to me from the narrative of Judah and Tamar and what happened to her, when simply read?"[205] These narratives are different from the narrative (*historiai*) about Samuel and the medium. Since the narrative in

---

[201]Trigg, "Eustathius of Antioch's Attack," 221.

[202]Origen, *Homily on 1 Kingdoms 28* 2.1 (Greer and Mitchell, 32); and ibid., 2.3 (Greer and Mitchell, 35).

[203]Ibid., 2.4 (Greer and Mitchell, 35), 9.7 (Greer and Mitchell, 59), and 10.2 (Greer and Mitchell, 61).

[204]Ibid., 2.1 (Greer and Mitchell, 33). The "elevated sense" (*anagōgēs*) here is another word that Origen uses for allegory, which "hints that a deeper meaning lies within the written word." See R. M. Grant, *The Letter and the Spirit* (London: SPCK, 1957), 124.

[205]Origen, *Homily on 1 Kingdoms 28* 2.2-3 (Greer and Mitchell, 35).

1 Samuel 28 "touches all people," its truth must be "in accordance with the word" (*kata ton logon*).[206]

The appeal Origen makes here between the elevated sense and that which is in accord with the letter is found in his well-known threefold sense of Scripture:

> One must therefore pourtray [*sic*] the meaning of the sacred writings in a threefold way upon one's own soul, so that the simple man may be edified by what we call the flesh of the scripture, this name being given to the obvious interpretation; while the man who has made some progress may be edified by its soul as it were; and the man who is perfect and like those mentioned by the Apostle: "We speak wisdom among the perfect; yet a wisdom not of this world, nor the rulers of this world, which are coming to naught; but we speak God's wisdom in a mystery, even the wisdom that hath been hidden, which God foreordained before the worlds unto our glory" [1 Cor 2:6-7] – this man may be edified by the spiritual law, which has "a shadow of the good things to come" [Heb 10:1]. For just as a man consists of body, soul and spirit, so in the same way does the scripture, which has been prepared by God to be given for man's salvation.[207]

This threefold sense is the basis of Origen's differentiation between the elevated and the narrative sense. In *On First Principles* he continues to explain that some passages do not have a fleshly sense at all, and those move us on to "seek only for the soul and the spirit, as it were, of the passage."[208] This is the broader interpretive context within which we must understand Origen when he asks, "What is the benefit for me from the narrative about Lot and his daughters? Likewise, what is the benefit to me from the narrative of Judah and Tamar and what happened to her, when simply read (Gen 38:1-30)?"[209]

---

[206]Ibid., 2.3 (Greer and Mitchell, 35).

[207]Origen, *On First Principles* 4.2.4 (Butterworth, 275-76). All quotations from *On First Principles* are from Origen, *On First Principles*, trans. G. W. Butterworth (Gloucester, MA: Peter Smith, 1973).

[208]Ibid. (Butterworth, 277-78).

[209]Origen, *Homily on 1 Kingdoms* 28 2.2-3 (Greer and Mitchell, 35). In *On First Principles* 4.2.2 (Butterworth, 272), Origen writes about the difficulty of understanding passages like "the intercourse of Lot with his daughters, or the two wives of Abraham, or the two sisters married to Jacob, or the two hand-maids that bore children by him."

There is certainly benefit derived from the fleshly sense, and multitudes of believers show this. But within that fleshly sense, "by means of stories of wars and the conquerors and the conquered certain secret truths are revealed to those who are capable of examining these narratives."[210] Thus, the intention of the author of Scripture, the Holy Spirit, was to make the bodily part of Scripture, which is "in many respects unprofitable," relevant and applicable to Christians.[211] It is narratives like Lot and his daughters and Judah and Tamar that Origen considers "unprofitable" in the fleshly sense.

Some narratives make good sense with regard to the obvious meaning, the fleshly sense. In these cases, there is no need to go deeper. However, some narratives contain certain "stumbling-blocks," which Origen describes as "hindrances and impossibilities to be inserted in the midst of the law and the history."[212] As he explains earlier in *On First Principles*, many Christians are confused about how to read some Old Testament Scriptures and "believe such things about [God] as would not be believed of the most savage and unjust of men."[213] The reason for this lack of understanding on the part of Jews and heretics, and the misunderstandings on the part of Christians, is because "scripture is not understood in its spiritual sense, but is interpreted according to the bare letter."[214] But these stumbling blocks are there as a kind of indicator to something deeper. Origen does not want the reader of these difficult passages to attribute anything to God that is unworthy of him.[215] In cases where Scripture appears to do this, Origen argues that this is warrant to move to the deeper, spiritual element.

One may be justified in thinking that Origen could see the narrative of the medium at Endor in these terms. That is, it would make sense,

---

[210]Origen, *On First Principles* 4.2.8 (Butterworth, 284-85).
[211]Ibid. (Butterworth, 285).
[212]Ibid., 4.2.9 (Butterworth, 285).
[213]Ibid., 4.2.1 (Butterworth, 269-71).
[214]Ibid., 4.2.2 (Butterworth, 271-72).
[215]For an excellent discussion on this issue, see Sheridan, *Language for God*, 27-44.

within Origen's own understanding, if he saw stumbling blocks in this narrative. Surely it is unworthy that the prophet Samuel was in Hades, that a medium was an appropriate intermediary between the human and divine worlds, and that magic is effective, especially on the holy people of God. The striking thing is, however, that Origen did not use these difficulties to avoid interpreting the passage "in accordance with the word." "For who," he asks, "once delivered from this life, wishes to be subject to the authority of a petty demon so that a belly-myther might bring up not just any believer but Samuel the prophet?"[216] Origen thus argues that the narrative must be taken on its own terms: "Its truth is necessary in accordance with the word."[217]

Origen knows that understanding the passage in this way has its difficulties. He confesses that if he concludes these things are not true, then it can induce people toward unbelief. But if he confesses they are true, we must investigate further because they are an occasion for doubt.[218] Origen also knows that there are some Christians who have "faced off against scripture" and claim that the narrative is not true. They conclude that the medium is lying, that Samuel was not brought up, and that Samuel does not speak.[219] Concluding, as Origen does, that Samuel really was the one who was brought up by the medium will require hard work and trial in scriptural interpretation, so he accepts the narrative as it stands and sets out to perform this hard work. For Origen, it is the Holy Spirit who is here speaking in the person (*prosōpon*) of the narrator, and he gives every indication that Saul "knew it was Samuel" (v. 14) and introduces his words with "Samuel said" (vv. 15-16) without reservation.[220]

---

[216]Origen, *Homily on 1 Kingdoms 28* 2.4 (Greer and Mitchell, 35).

[217]This does not mean that Origen believes the text has no elevated sense, however. Origen returns to the elevated sense later in his treatise. See Margaret M. Mitchell, "Patristic Rhetoric on Allegory: Origen and Eustathius Put 1 Kingdoms 28 on Trial," *Journal of Religion* 85, no. 3 (2005): 423, reprinted in Greer and Mitchell, *"Belly-Myther" of Endor*.

[218]Origen, *Homily on 1 Kingdoms 28* 2.5 (Greer and Mitchell, 37).

[219]Ibid., 3.1 (Greer and Mitchell, 37).

[220]Trigg, "Eustathius of Antioch's Attack," 226.

See what a great trial there is in God's word, which requires hearers able to listen to words that are holy, great, and ineffable, words concerning our departure from this world. . . . Indeed, the passage needs further examination, and I say that both the narrative sense and the examination of this passage are necessary so that we may discern what our condition will be after we depart from this life.[221]

It thus makes sense that Origen's first order of business in his actual exegesis is to remove obstacles keeping us from reading "the words that stand written."[222] These are, after all, what the Holy Spirit in the persona of the narrative wrote.[223] He thus proceeds to give reasons why it should not disturb the Christian that Samuel was in Hades.[224] Thus, even though Samuel did come up from Hades to converse with Saul, this should not be problematic for believers. As Origen himself states, "Therefore, there is no stumbling block in this passage."[225]

EUSTATHIUS'S APPROACH TO THE PASSAGE

As we have seen, Eustathius accuses Origen not of allegory but of being literal in his exegesis of 1 Samuel 28. A literal reading, like Origen's, would have to accept that a medium raised a holy prophet up from the dead to speak with Saul. Eustathius denied that Samuel actually appeared, because this interpretation would encourage Christians to resort to magic and necromancy, disparage the power of God (who alone can raise the dead), and undermine Christian morality by implying that the just and the unjust have the same ultimate destiny.[226]

---

[221]Origen, *Homily on 1 Kingdoms 28* 4.10–5.1 (Greer and Mitchell, 43).

[222]Ibid., 4.2 (Greer and Mitchell, 39).

[223]Ibid.: "What in fact are the words that stand written? 'And the woman said, "Whom shall I bring up for you?"' (I Kgdms 28:11). Whose persona is it that says, 'the woman said'? Is it, then, the persona of the Holy Spirit by whom scripture is believed to have been written, or is it the persona of someone else? For, as those who are familiar with all sorts of writings know, the narrative persona throughout is the persona of the author. And the author responsible for these words is believed to be not a human being, but the author is the Holy Spirit who has moved the human beings to write."

[224]Ibid., 4–9 (Greer and Mitchell, 39-59). Trigg ("Eustathius of Antioch's Attack," 227-28) provides a helpful summary of the actual exegesis.

[225]Ibid., 10.1 (Greer and Mitchell, 59).

[226]The reasons given here are summarized from Trigg, "Eustathius of Antioch's Attack," 229.

Eustathius's conclusion was that Saul was deceived by a demon posing as Samuel.[227]

Eustathius looks for a warrant to explain why the reader should not accept the words as they stand. For example, when 1 Samuel 28:12 states that "the woman saw Samuel," and not a demon posing as Samuel, Eustathius has recourse to Exodus 7:11-12, which appears to claim that "the deeds of the sorcerers are similar to those accomplished by Moses."[228] For Eustathius, Scripture does not intend to say that sorcery did the same as that accomplished by the divine.[229] Rather, "What was accomplished by Moses displayed the truth by what was done, but what was stitched together by the magicians conjured up images in appearance only."[230] In this and other scriptural narratives, Scripture assumes that the readers understand the magic to be delusive since it is clearly stated beforehand that the story is about sorcery, "and every one knows that sorcery is just daemonic deceit."[231] In the words of Eustathius:

> The author, by setting forth from the outset the name of "the magicians" and the devices "of sorcery" (Exod 7:10), has demonstrated that there would be no need at all to dispute that fact that each of these deeds was accomplished by magicians using the art of magic.
>
> The dramatic actions of the narrative we are speaking about are set in writing in just this way also. For by naming her "belly-myther" and then by reproving the impious devices of divination and showing that the persona seeking divination is possessed by a demon, the narrator has made it absolutely clear that each of these acts was accomplished by spurious illusions. This is the manner in which these elements of the narrative are to be understood.[232]

See also Eustathius, *On the Belly-Myther* 3.4 (Greer and Mitchell, 69); ibid., 3.3 (Greer and Mitchell, 67-69); and ibid., 14.7 (Greer and Mitchell, 107-9).

[227]Eustathius, *On the Belly-Myther* 3-4 (Greer and Mitchell, 69-75).

[228]Ex 7:11-12: "Then Pharaoh summoned the wise men and the sorcerers; and they also, the magicians of Egypt, did the same by their secret arts. Each one threw down his staff, and they became snakes; but Aaron's staff swallowed up theirs." See also Eustathius, *On the Belly-Myther* 9.3 (Greer and Mitchell, 85).

[229]Eustathius, *On the Belly-Myther* 9.4 (Greer and Mitchell, 85). In response to this possibility, Eustathius exclaims, "Surely not!"

[230]Ibid.

[231]Smelik, "Witch of Endor," 167.

[232]Eustathius, *On the Belly-Myther* 9.13 (Greer and Mitchell, 87-89).

ANALYSIS

What are we to make of this example? It is apparent that the traditional distinction between Antioch and Alexandria along the lines of literal/historical versus allegorical reading cannot be applied here, at least in the way it has been traditionally understood. Origen and Eustathius actually had some important things in common in their understandings of biblical interpretation. Both are convinced of Scripture's inspiration by the Holy Spirit. They were also both convinced of Scripture's contemporary relevance to the Christian. The difference between them lay in Eustathius's challenge to Origen's application of Scripture. From our perspective, both interpretations were anachronistic in the sense that they paid little attention to the author's intended meaning. Since Scripture was inspired by the Holy Spirit, he was seen as the true author, and this broke down the historical anchoring and distancing common in our modern approaches.

But there were also some important differences between the approach of Eustathius and that of Origen. The most significant and far-reaching difference can be seen in Eustathius's criticism of Origen for focusing "attention on 'names' [*onomasi*] and not on facts [*pragmasi*]."[233] Both men deduced conclusions from the text.[234] For Eustathius, one of the main aims of the treatise is to discredit the testimony of the medium. He thus makes a "nontextually supported assumption" that the divinatory arts are the work of demons.[235] In fact, this is self-evident to both Eustathius and Origen even though it is not explicitly indicated in the scriptural passage. This is "one clear indication of the 'literal' limits of both interpretations."[236] For Eustathius, a literal reading of the text is guided by his initial deduction about the medium. But as he progresses in his counter-interpretation, he explains that the medium seduced Saul

---

[233]Ibid., 1.3 (Greer and Mitchell, 63).

[234]Young, *Biblical Exegesis*, 163.

[235]Mitchell, "Patristic Rhetoric," 439. This is latent throughout the treatise. Examples include 3.3, 10.1, 11.1, 20.2.

[236]Mitchell, "Patristic Rhetoric," 439.

into believing that he had actually seen Samuel.[237] This is a reading that respects the narrative coherence of the story. Origen's reading, according to Eustathius, does not do this and is, therefore, invalidated in his eyes.

Eustathius's approach is what we have earlier called "ikonic," while Origen's is "symbolic."[238] The criticism that Origen focused on "names" rather than "facts" must be understood in this light. Eustathius shows that his concern is with Origen's treatment of words as tokens or symbols (names) rather than an approach that has the narrative coherence (facts) of the passage at center.[239] This is why Eustathius's entire document is "a series of rationalistic arguments, bolstered by scriptural parallels, to prove Origen's lexical reading of the text is on the wrong lines, and so everything he deduces from it is unacceptable."[240] Eustathius's understanding of the whole passage mitigates against Origen's view. There is no statement in the passage stating explicitly that Saul actually saw Samuel – he only thinks he did.[241]

Interestingly, Eustathius's rationalistic arguments and scriptural proofs are less literal than Origen, but offer a more satisfactory conclusion.[242] Indeed, at many places in the treatise, Eustathius gives examples of other scriptural narratives where commonsense inference is required because certain details are not spelled out in the text. This serves to justify his nonliteral interpretation.[243]

We see here that Eustathius is not concerned with any such allegorical method in Origen's exegesis but rather with the doctrine he draws from the text. He does not think that Origen respects the sequence of the story,

---

[237]Eustathius, *On the Belly-Myther* 7 (Greer and Mitchell, 79-81).

[238]See above, chapter three, under "*Mimēsis.*" In the words of Frances Young: "The 'ikon' will resemble the person or event which it represents, but symbols are not representations in that sense; symbols are 'tokens' or 'signs' whose analogous relationship with what is symbolized is less clear. Indeed, whereas a text offers clues to its 'ikonic' intention, the clues to 'symbolic' language lie in the impossibilities which cannot be taken at face value." See Young, *Biblical Exegesis*, 210.

[239]Young, *Biblical Exegesis*, 163.

[240]Ibid.

[241]Young, "Rhetorical Schools," 194.

[242]Ibid.

[243]Young, *Biblical Exegesis*, 164.

the intention of the author, or the coherence of the narrative with the rest of Scripture.[244] Eustathius takes every detail of the conversation between the medium and Saul and attempts to show that a demon was conjured up rather than Samuel and that the narrative makes sense in no other way. As further confirmation of this reading, references to scriptural laws against sorcery and consulting mediums are presented.[245]

It was ultimately the sequence of the story, both internally and with the rest of Scripture, that was of concern to Eustathius, and it was this against which Origen transgressed. Even the word *engastrimythos* — translated here as "belly-myther," "medium," or "witch" — is taken by Eustathius to show that the author intended to imply that the medium was false. Origen is chastised for not recognizing this.[246]

Eustathius wants to prove that the medium is speaking falsehoods, and in the process he resorts to standard rhetorical techniques discussed earlier in this chapter. In applying this standard exegetical practice, he is actually not far from allegorization himself.[247] For example, he makes specific reference to "the rhetorical handbooks," which show that "a myth is a fabrication composed with persuasive attraction with an eye to some matter of vital importance at utility."[248] A myth is a "fabrication"; it "fashions in speech a likeness of concrete events, though it is bereft of fact."[249] The aim of myth is to persuasively show that "what does not exist does exist, and it introduces in narrative form a fabricated copy."[250] Just as a painting represents or imitates (*mimēsis*) its setting, so does myth.

He continues with an appeal to Plato, quoting his *Republic* fairly extensively to explain that myths are actually a way of educating children. Initially, children are enchanted by myths, but eventually they are

[244]Young, "Rhetorical Schools," 194.
[245]Young, *Biblical Exegesis*, 164.
[246]Eustathius, *On the Belly-Myther* 26.9-10 (Greer and Mitchell, 149).
[247]Mitchell, "Patristic Rhetoric," 439.
[248]Eustathius, *On the Belly-Myther* 27.2 (Greer and Mitchell, 149-51).
[249]Ibid., 27.2 (Greer and Mitchell, 151).
[250]Ibid., 27.3 (Greer and Mitchell, 151).

taught to distinguish between truth and fiction.[251] Eustathius explains that the poems of Homer and Hesiod have a pride of place in the "school curriculum."[252] Plato has described these poems of Homer and Hesiod to be false myths, so "how much more," claims Eustathius, "will one designate the words of the demon-colluding old crone uttered as false myth-making, when even the name belly-myther includes this meaning?"[253] In what can be read as a slight against Origen, Eustathius claims that even school children can recognize this. Thus,

> the techniques of narrative criticism are no doubt in his mind. Eustathius' purpose is to prove his basic point about the *engastrimythos*, but in the process he uses standard etymological methods, explicitly refers to literary-critical observations about art being a *mimēsis* of life, makes a learned reference to Plato as any well-educated teacher would, and actually mentions rhetorical textbooks. His interpretation is rooted in the traditional *paideia* of the rhetorical schools.[254]

By connecting the belly-myther to myth in this way, Eustathius has established that the witch of Endor is untrustworthy. Thus, guided by his initial deduction about her, and since she "says things other than what they are," his literal reading has connected her with Origen because Origen believes her.[255] The accusation that the medium was "saying things other than what is actually the case, allegedly pronouncing the names of perceptible things and indicating spiritual matters," is tantamount to calling her an allegorist![256] Since Origen believes her and therefore promotes her, he falls into the same criticism. Even though Origen is actually the one with a more literalist interpretation of the passage, "Eustathius brilliantly executes a sharp denunciation of Origenic allegorical method as being in collusion with such a 'belly-mythologist.'"[257]

---

[251]Ibid., 28 (Greer and Mitchell, 153-55).
[252]Ibid., 29.1 (Greer and Mitchell, 155).
[253]Ibid., 29.2 (Greer and Mitchell, 155).
[254]Young, *Biblical Exegesis*, 176.
[255]Eustathius, *On the Belly-Myther* 6.9 (Greer and Mitchell, 79).
[256]Mitchell, "Patristic Rhetoric," 438.
[257]Ibid.

Although Origen insists that his interpretation of 1 Samuel 28 is according to the letter, Eustathius makes him into a spokesperson for allegory. Because Origen takes the words of the belly-myther literally – that is, not as myth – he is practicing the very thing that Eustathius rejects at the outset. For Eustathius, Origen "does what she does, she does what he does: pull words out of the air to say that what isn't there actually is."[258]

Ultimately Eustathius insists on looking for the *historia* and the *gramma* of the text. As we have seen, this is not history in the modern sense but more like the narrative logic of the text, or a coherent account of things that does not have recourse to merely the verbal details of the text, as Origen does. Origen, states Eustathius, "piles up so many passages that are in no way relevant to the inquiry at hand, he passes over passages that by their very nature are related."[259] Young affirms that Eustathius "was simply using standard literary techniques deriving from the treatment of texts in the rhetorical schools to protest against esoteric philosophical deductions being made in what he regarded as an arbitrary way."[260] As we have seen, both made deductions from the text; however, the focus against Origen was his tokenist rather than his allegorical exegesis. Origen, in Eustathius's thinking, did not pay enough attention to the coherence of the narrative.[261] Allegory was a standard figure of speech in the ancient world as it is today.[262] If a text carried clear indications of its presence, it could be allowed. Thus, it was not necessarily allegory that Eustathius objected to but rather the type of allegory that destroyed textual coherence.

We may appear to have gone much too far afield in this chapter from our main concern in the book of how the church fathers understood

---

[258]Ibid., 439. See Eustathius, *On the Belly-Myther* 6.9: The medium was "saying things other than what is really the case, allegedly pronouncing the names of perceptible things and indicating spiritual matters."

[259]Eustathius, *On the Belly-Myther* 26.1 (Greer and Mitchell, 145-47).

[260]Young, "Rhetorical Schools," 195.

[261]Young, *Biblical Exegesis*, 164.

[262]See O'Keefe and Reno, *Sanctified Vision*, 89-93.

Genesis 1. Doesn't this background in patristic hermeneutics serve to confuse the issue? I would answer with a forceful no. Part of the point I am trying to make is that we cannot simply parachute into the context of the Fathers and disregard it by plucking out quotations that appear to support our conclusions. Great care is required to understand the world into which we enter, and this entails addressing some foundational issues. When modern assumptions about biblical interpretation are projected onto the Fathers, we run the risk of making them champions of some idea or concept that they simply were not. To assume that the Fathers read "literally" in the same way we mean "literal" is a misrepresentation of their context. It needs to be corrected if we are to move forward and speak meaningfully about how the Fathers understood Genesis 1.

RECOMMENDED READING

Louth, Andrew. *Discerning the Mystery: An Essay on the Nature of Theology.* Oxford: Clarendon Press, 2003.

O'Keefe, John J. "'A Letter That Killeth': Toward a Reassessment of Antiochene Exegesis, or Diodore, Theodore, and Theodoret on the Psalms." *Journal of Early Christian Studies* 8, no. 1 (2000): 83-103.

O'Keefe, John, and R. R. Reno. *Sanctified Vision: An Introduction to Early Christian Interpretation of the Bible.* Baltimore, MD: Johns Hopkins University Press, 2005.

Young, Frances M. *Biblical Exegesis and the Formation of Christian Culture.* Peabody, MA: Hendrickson, 2002.

PART II

READING

the

FATHERS

4

BASIL THE LITERALIST?

Therefore, let it be understood as it has been written.

BASIL, HEXAEMERON 9.1 (FC 46:136)

The previous chapter on biblical interpretation in the early church has provided the necessary foundation required to answer the question posed in this chapter – Is Basil a literalist? We will see that a patient and deliberate approach to this question resists the temptation to make rash conclusions drawn from proof texts.

This chapter's epigraph from Basil of Caesarea (329–379) is contained in his ninth homily on the six days of creation, known in the early church as the Hexaemeron.[1] The words are spoken in the context of allegory. Basil explains that he knows all about the rules of allegory, but he apparently proceeds to reject it in favor of the "common meaning," hence the call to his hearers to understand Genesis "as it has been written."

The extended section, which will be cited in full below, has sometimes been used as a hammer to prove the great Cappadocian's utter

---

[1] All *Hexaemeron* quotations are from Saint Basil, *Exegetic Homilies*, trans. Agnes Clare Way, FC 46 (Washington, DC: Catholic University of America Press, 1963).

rejection of an allegorical or spiritual interpretation of Genesis 1 in particular and Scripture in general.[2] Once this rejection is established, Basil is then sometimes adopted as a literalist and a champion of creationism in the same vein as groups like Answers in Genesis and Creation Ministries International.[3]

For example, in "Genesis Means What It Says: Basil (AD 329–379)," the anonymous author uses Basil to debunk the apparent misconception that "Christians in the early Church took a more allegorical view of things."[4] Basil is identified as a key figure in the fourth-century church who defended the doctrine of the Trinity against the Arians. His "classic Trinitarian formula" (three persons, one substance) is "still one of the best summaries of the Biblical doctrine, and is accepted by all branches of orthodox Christianity."[5] Basil is significant for these theological contributions in addition to being a pastor greatly respected and admired for his establishment of charitable institutions, hospitals, and schools and organization of famine relief. In the Eastern Orthodox Church, he is considered a saint. It is important for the author of this article to establish the significance and trustworthiness of this early Christian "pastor and church leader" because his homilies on the six days of creation (Hexaemeron) are then used to prove that the early church did not, in fact, take "a more allegorical view of things." In other words, Basil's *Hexaemeron* is wielded in support of a literalist understanding of Genesis.[6]

---

[2]As we will see below, *Hexaemeron* 3.9 is also offered as proof of Basil's rejection of allegory and, therefore, acceptance of literalism.

[3]See https://answersingenesis.org/ and http://creation.com/.

[4]Anonymous, "Genesis Means What It Says: Basil (AD 329–379)," accessed September 30, 2016, http://creation.com/genesis-means-what-it-says-basil-ad-329-379. Identification of this article's author is elusive. According to the article's first footnote, it was adapted from David Watson, "An Early View of Genesis One," *Creation Research Society Quarterly* 27, no. 4 (1991): 138-39. The anonymous version also appears at https://answersingenesis.org/genesis/genesis-means-what-it-says-basil-ad-329-379/.

[5]Anonymous, "Genesis Means What It Says."

[6]As will be seen below, I make a contextual distinction between "literal" as understood in the patristic age and "literal" as understood by the author of this article and others like it. Thus, the choice of "literalist" here is deliberate to convey a difference that will be explained below.

The remainder of the anonymous article proceeds by using proof texts from the *Hexaemeron* to bolster the already accepted notion that Basil was a literalist, including the key passage found at the beginning of this chapter from *Hexaemeron* 9.1. No commentary or explanation on the passage is offered – apparently it is clear that Basil was a literalist.

Another example describing Basil in nearly identical terms in support of literalism is found in James R. Mook's article "The Church Fathers on Genesis, the Flood, and the Age of the Earth."[7] Similarly, with the authority of Basil established, Mook simply proof-texts a pastiche of statements from Basil's *Hexaemeron* 9.1 that are intended to prove the already stated conclusion that Basil was a literalist: "Basil of Caesarea specifically opposed the 'distorted meaning of allegory,' accusing allegorists of serving 'their own ends' and giving 'a majesty of their own invention to Scripture'; advocating instead a humble acceptance of the 'common sense,' the 'literal sense' of Scripture 'as it has been written.'"[8]

The two examples above are quite certain that Basil's words indicate an outright rejection of allegory and, perhaps more importantly, an explicit acceptance of what they call "literal" interpretation. But is this a fair reading of the great Cappadocian? This chapter sets out to answer that question.

BASIL'S *HEXAEMERON*

Basil's *Hexaemeron* is a series of nine homilies on the six days of creation (Gen 1:1–2:3) that were most likely delivered during Lent.[9] The date of

---

[7]James R. Mook, "The Church Fathers on Genesis, the Flood, and the Age of the Earth," in *Coming to Grips with Genesis: Biblical Authority and the Age of the Earth*, ed. Terry Mortenson and Thane H. Ury (Green Forest, AZ: Master Books, 2008), 23-52, esp. 30-32.

[8]Ibid., 30. Mook employs the translation by Jackson found in NPNF[2] vol. 8.

[9]Hughes Old acknowledges that while most affirm these homilies were preached at Lent, he believes this may not be the case: "There is a reference to fasting at the end of the eighth homily, but this may refer to Friday fasting rather than Lenten fasting. Certainly the theme of the homily does not suggest Lent, nor is there any indication that they were preached to catechumens." Hughes Oliphant Old, *The Reading and Preaching of the Scriptures in the Worship of the Christian Church*, vol. 2, *The Patristic Age* (Grand Rapids: Eerdmans, 1998), 43.

delivery is difficult to attain. There is agreement that they belong to the latter part of Basil's tenure as bishop of Caesarea (370–379), probably sometime around the middle of 378.[10] They were delivered extemporaneously, without a prepared manuscript, over a period of five days.[11] Thus, the first four homilies were delivered in the mornings and evenings of the first two days, the fifth delivered the next day, and the last four delivered on the final two days in the mornings and evenings. The homilies were recorded by a stenographer and published with very little editing from Basil.[12]

Basil's nine homilies on the Hexaemeron were an acknowledged masterpiece very quickly after they became available. Basil's fellow Cappadocian and friend Gregory of Nazianzus states,

> When I take his *Hexaemeron* in my hand and read it aloud, I am with my Creator, I understand the reasons for creation more than I formerly did when I used sight alone as my teacher. . . . When I read his other explanations of Scripture, which he unfolds for those who understand but little, writing in a threefold manner on the solid tablets of my heart, I am prevailed upon not to stop at the letter, nor to view only the higher things, but to pass beyond and to advance from depth to depth, calling upon abyss through abyss, finding light through light, until I reach the loftiest heights.[13]

Basil's brother, Gregory of Nyssa, had even higher praise for the homilies: "Furthermore, we have access to that divinely inspired

---

[10]McGuckin argues that Basil delivered them as a priest during Lent sometime between 365–370. See John A. McGuckin, "Patterns of Biblical Exegesis in the Cappadocian Fathers: Basil the Great, Gregory the Theologian, and Gregory of Nyssa," in *Orthodox and Wesleyan Scriptural Understanding and Practice*, ed. S. T. Kimbrough (Crestwood, NY: St. Vladimir's Seminary Press, 2005), 45.

[11]Richard Lim, "The Politics of Interpretation in Basil of Caesarea's 'Hexaemeron,'" *Vigiliae Christianae* 44, no. 4 (1990): 351; Andrew Louth, "The Six Days of Creation According to the Greek Fathers," in *Reading Genesis After Darwin*, ed. Stephen C. Barton and David Wilkerson (Oxford: Oxford University Press, 2009), 44; and Isabella Sandwell, "How to Teach Genesis 1.1-19: John Chrysostom and Basil of Caesarea on the Creation of the World," *Journal of Early Christian Studies* 19, no. 4 (2011): 542.

[12]Old, *Reading and Preaching of the Scriptures*, 2:43. The extemporaneous preaching and its recording as is follows the pattern of classical rhetoric. Basil's *Hexaemeron* is a very good example of a homily that employs rhetoric as a tool.

[13]Gregory of Nazianzus, *Homily* 43, cited from Agnes Clare Way, introduction to *Saint Basil: Exegetic Homilies*, FC 46 (Washington, DC: Catholic University of America Press), vii-viii.

[*theopneuston*] study of our father [Basil the Great] whose exposition everyone treasures as not being inferior to what Moses had taught."[14]

While we might agree that the words of the two Gregorys above may be somewhat hyperbolic, we must also give a nod to the esteem and influence of Basil's *Hexaemeron*. It has been described as having utmost importance and is well known for being the foundation for the later *Hexaemera* of Ambrose, pseudo-Eustathius, Philoponus, and Glyca.[15] Indeed, these later *Hexaemera* were "little but revampings of Basil's work."[16] Ambrose even translated into Latin some passages from Basil's Greek work and inserted them directly into his own *Hexaemeron*.[17] Further, Rufinus and Eustathius both made Latin translations of the work.[18]

AUDIENCE

There are a few passages in the homilies that relate information about the makeup of Basil's hearers. First among these is Homily 3.1:

> It has not escaped my notice, however, that many workers of handicrafts, who with difficulty provide a livelihood for themselves from their daily toil, are gathered around us. These compel us to cut short our discourse in order that they may not be drawn away too long from their work.

From this we see that some tradespeople actually took the day off, or a portion thereof, to hear Basil preach. These were skilled workers who worked hard for a living.

In Homily 7.5-6 Basil chastises abusive husbands and those who are morally lax. In his final statement in Homily 7 he encourages his congregation to discuss the homily he had already delivered and to fall asleep at night "while engaged in the thought of these things."[19] Near

---

[14]Gregory of Nyssa, *Hexaemeron*, accessed September 30, 2017, www.lectio-divina.org/index .cfm?fuseaction=feature.display&feature_id=367.
[15]Frank Egleston Robbins, *The Hexaemeral Literature: A Study of the Greek and Latin Commentaries on Genesis* (Chicago: University of Chicago Press, 1912), 42.
[16]Ibid.
[17]Way, *Saint Basil*, viii.
[18]Ibid.
[19]Basil *Hexaemeron* 7.6 (FC 46:116).

the end of Homily 8, realizing that he is going "beyond due limits," he exhorts the congregation to care for the soul and leave the pleasures of the flesh. He fears, however, that this will be ignored by some who will merely return to the gaming table.[20]

There also exists a helpful description of Basil's audience from his brother, Gregory of Nyssa:

> Among the many listeners were some who grasped his loftier words, whereas others could not follow the more subtle train of his thought. Here were people involved with private affairs, skilled craftsmen, women not trained in such matters together with youths with time on their hands; all were captivated by his words, were easily persuaded, led by visible creation and guided to know the Creator of all things. Should anyone assess the words intended by the great teacher, no doubt he would not omit a single one.[21]

From all this, some scholars conclude that Basil's audience for these homilies consisted of the proverbial man on the street – those who were not necessarily spiritually advanced Christians.[22] We must be careful, however, to avoid characterizing the audience as only "blue collar" people. Isabella Sandwell's balanced conclusion about the audience is well stated: "It is thus likely that Basil's audience . . . was a mixed, urban audience containing some members of the educated elite but also many ordinary craftsmen, women, and younger people."[23]

HOW DOES BASIL READ SCRIPTURE?

In addition to the important general clarifications made in the previous chapter about early Christian biblical interpretation, we do well to consider Basil's own broader interpretive context through some of his other, non-Hexaemeral writings. Can we simply transfer our own modern understanding of "literal" and "allegorical" onto Basil and the

---

[20]Ibid., 8.8 (FC 46:133).
[21]Gregory of Nyssa, *Hexaemeron.*
[22]Lim, "Politics of Interpretation," 361.
[23]Sandwell, "How to Teach Genesis 1.1-9," 542.

early church? Are these terms self-evident in meaning? A closer look at Basil's interpretive context will serve to highlight the difficulty of this direct transference.

Mook illustrates the problem of direct transference for us. In "The Church Fathers on Genesis," Mook appropriately gives an argument for the importance of the church fathers to orthodox Christianity because of the significant role they played in theological debate and, thus, in clarifying the parameters of orthodoxy.[24] He continues by exhorting evangelicals to also consider the importance of the Fathers through the renewal of interest in them that is occurring today. Of course, Mook is correct. The flowering of interest in the church fathers among evangelicals is at a stage never before seen. Many, including myself, agree with Mook that we would do well to at least understand their importance to the Christian tradition. But Mook's appeal also includes a word of caution:

> Since their voice on theological matters has always been coveted, it would be expected that, along with cautious use of their wisdom, there is also a tendency with some to misread the patristic literature. The teachings of the fathers can just as surely be taken out of context, eisegeted, or muffled altogether, as the Scriptures can be.[25]

This then forms the purpose of Mook's article – to point out and counter what he sees as misreadings of the Fathers by some who use them to argue against the creation science perspective.

In an explicit appeal for his readers to be serious about our own Christian heritage, Mook offers this fine piece of advice:

> Christians should be aware of the great cloud of witnesses in Church history, and a judicious use of the fathers can be both relevant and edifying. And even though the Christian's highest and final authority should always be Scripture, the more knowledge of Church history one has, the better. In being tutored by the fathers, we will be better armed to discern and respond to the novel theological heterodoxies in their day and ours.[26]

---

[24]Mook, "Church Fathers on Genesis," 24.
[25]Ibid.
[26]Ibid., 25.

In keeping with his stated purpose, Mook takes to task authors who, in his opinion, misread the Fathers in support of the "day-age view" and "framework hypothesis."[27] While my purpose in this chapter is not to deal with these specific issues, it is Mook's critique of these authors, based as it is in his apparently correct reading of the Fathers, that is of concern to me. In other words, I am concerned with just how accurate Mook's own understanding of Basil really is.

Mook accuses authors like Hugh Ross and Gleason Archer of misreading the Fathers.[28] But does Mook actually read them properly? He makes a fundamental distinction among the church fathers between the "literalists" and the "allegorists." According to him, "In the ancient Church there was a tension between allegorists and literal interpreters."[29] This tension is really presented as an opposition. He then proceeds to set these literalists in opposition to the allegorists, with the general conclusion that the literalists read the Bible properly whereas the allegorists read it improperly. In other words, the literalists are biblical and the allegorists are not. I have addressed the general issue in the previous chapter with the conclusion that Mook's distinction is simplistic to the point of misrepresenting the interpretive context in which the Fathers lived and wrote. Here I will test Mook's distinction with Basil's approach to biblical interpretation.

Many scholars of the church fathers find Basil's comments on allegory less than clear and obvious. In the introduction to the first English translation of Basil's *Hexaemeron*, Blomfield Jackson notes that "it is an innovation for Basil to adopt such an exclusively literal system of exposition as he does, – *e.g.* in Hom. IX. on the Hexæmeron, – the system which is one of his distinguishing characteristics."[30] Peter

---

[27]The day-age view holds that the days in Gen 1 are longer than twenty-four hours. The framework hypothesis is a view that examines the literary structure of Gen 1:1–2:4 to show that the purpose of the passage may not be historical.

[28]Hugh Ross and Gleason Archer, "The Day-Age View," in *The Genesis Debate: Three Views on the Days of Creation*, ed. David G. Hagopian (Mission Viejo, CA: Crux Press, 2001), 123-63.

[29]Mook, "Church Fathers on Genesis," 29.

[30]Blomfield Jackson, "Introduction" (NPNF[2] 8:xlv).

Bouteneff urges readers to consider not only the larger context of Basil's *Hexaemeron* but also his larger literary output: "To interpret this passage as anti-allegorical either makes Basil sound positively hypocritical, or misrepresents him. For elsewhere he either acknowledges multiple senses of Scripture or is seen making extensive (if sometimes only implicit) allegorical use of it."[31] Indeed, it is reasonable to wonder about the tenor of Basil's comments on allegory in light of his approach to Scripture elsewhere in the *Hexaemeron* and some of his other writings.

LITERAL AND ALLEGORICAL IN BASIL'S NON-HEXAEMERAL WORKS

In what is perhaps Basil's best-known and most influential work, *On the Holy Spirit*, he offers a defense of the doctrine of the Holy Spirit.[32] It was written fairly late in his life and episcopacy[33] and thus is a product of his mature preaching.[34] Basil had received some criticism for introducing a doxology into common prayers that was more in common with the practice of Western churches.[35] One major objection of Basil's critics was that the doxology he introduced – "Glory to the Father, and to the Son, and to the Holy Spirit" – was an innovation. The critics

---

[31]Peter C. Bouteneff, *Beginnings: Ancient Christian Readings of the Biblical Creation Narratives* (Grand Rapids: Baker Academic, 2008), 129. I am reminded here of Philip Rousseau's comment about Basil's argument in his *Hexaemeron* that texts had to be taken at face value, thus taking words according to their "common use." Of this, Rousseau states, "It is probably fair to say that Basil promptly failed to follow his own advice." See Philip Rousseau, *Basil of Caesarea* (Berkeley: University of California Press, 1998), 322-23.

[32]The importance of this treatise in the early church is without doubt. Pelikan states that "what Basil and his contemporaries worked out as orthodox teaching [on the Holy Spirit] is what most of Christendom has gone on confessing." See Jaroslav Pelikan, "The 'Spiritual Sense' of Scripture: The Exegetical Basis for St. Basil's Doctrine of the Holy Spirit," in *Basil of Caesarea: Christian, Humanist, Ascetic*, ed. Paul Jonathan Fedwick (Toronto: Pontifical Institute of Medieval Studies, 1981), 2:338.

[33]Basil died on January 1 or 2, 379. Pelikan dates *On the Holy Spirit* sometime between the end of 374 and the end of 375. See Pelikan, "'Spiritual Sense' of Scripture," 338.

[34]McGuckin, "Patterns of Biblical Exegesis," 46. This is a significant point because it challenges the argument that Basil's move to literalism occurred later in life – before he preached on the *Hexaemeron*. This document is written after his *Hexaemeron*.

[35]Customary in Eastern churches was the doxology "Glory to the Father, through the Son, in the Holy Spirit." Basil replaced this with "Glory to the Father, and to the Son, and to the Holy Spirit." See McGuckin, "Patterns of Biblical Exegesis," 46.

argued that this expression of coequality between Father, Son, and Spirit had no explicit scriptural support.[36] In response to their insistence for "written proof," Basil recognized that it was not just about the words but rather about how they were used to convey their message.[37]

Jaroslav Pelikan has aptly summarized the issue involved here as it relates to the literal/allegorical issue in Basil, and sets out to analyze the "spiritual sense" of Scripture in *On the Holy Spirit*.[38] The Fathers built on the antithesis established by Paul between letter and spirit in 2 Corinthians 3, using it as justification for typological and allegorical interpretation of the Old Testament. Basil himself made an opposition between the ministry of the law and the ministry of the Holy Spirit based on this chapter.[39] In Moses, there was no "spiritual grace," only the law.[40] Only the person who has the Spirit of God is led by the Spirit.

> For of the senses in which "*in*" is used, we find that all help our conceptions of the Spirit. *Form* is said to be *in Matter*; *Power* to be *in* what is capable of it; *Habit* to be *in* him who is affected by it; and so on. Therefore, inasmuch as the Holy Spirit perfects rational beings, completing their excellence, He is analogous to Form. For he, who no longer "lives after the flesh," [Rom 8:12] but, being "led by the Spirit of God," [Rom 8:14] is called a Son of God, being "conformed to the image of the Son of God," [Rom 8:29] is described as spiritual.[41]

---

[36]Basil, *On the Holy Spirit* 27.68 (NPNF² 8:43). "Yet they never stop dinning in our ears that the ascription of glory '*with*' the Holy Spirit is unauthorized and unscriptural and the like." Italics in original translation.

[37]Ibid., 10.25 (NPNF² 8:16).

[38]The summary that follows of this book draws from Pelikan, "'Spiritual Sense' of Scripture," 338-40.

[39]Basil, *On the Holy Spirit* 24.55 (NPNF² 8:35). "According to the Apostle there is a certain glory of sun and moon and stars [1 Cor 15:41], and 'the ministration of condemnation is glorious' [2 Cor 3:9]. While then so many things are glorified, do you wish the Spirit alone of all things to be unglorified? Yet the Apostle says 'the ministration of the Spirit is glorious' [2 Cor 3:8]. How then can He Himself be unworthy of glory?"

[40]Basil, *On the Holy Spirit* 14.32 (NPNF² 8:20). "What spiritual gift is there through Moses? What dying of sins is there?"

[41]Ibid., 26.61 (NPNF² 8:38). Italics in original translation.

Since Scripture was "written through the inspiration of the Spirit," it was thus necessary that a "spiritual sense" was often conveyed by means derived from physical objects.[42]

Pelikan explains that "it was precisely this 'Spiritual sense' of scripture that was in contention between Basil and his critics."[43] Basil's theology insisted on the equality in being of the Father, Son, and Spirit. He knew that there was no explicit biblical support for his doxological "innovation" that advocates a coequality of the three. But, insists John McGuckin, "this . . . is not the same as having no support in the Apostolic Tradition, because the sense of the scripture (included in the Apostolic Tradition) is not reducible to its literal grammar."[44] Basil confirms this: "But we do not rest only on the fact that such is the tradition of the Fathers; for they too followed the *sense* of Scripture."[45]

Basil is quite clear about the danger of an overly literal reading of Scripture, explaining why he believes this near the beginning of the treatise: "The petty exactitude of these men about syllables and words is not, as might be supposed, simple and straightforward; nor is the mischief to which it tends a small one. There is involved a deep and covert design against true religion."[46] Thus, even though these critics wanted biblical language and only explicit biblical teaching included in Basil's doxology, he argues that they were actually battling against Scripture.[47]

In a section of *On the Holy Spirit* that explicitly deals with Paul's antithesis between law and spirit in 1 Corinthians 3, Basil advocates for an approach to Scripture that is sensitive to the Spirit. Referring to

---

[42]Ibid., 21.52 (NPNF[2] 8:34). See also Basil, *On the Holy Spirit* 26.62 (NPNF[2] 8:39). "For words applicable to the body are, for the sake of clearness, frequently transferred in scripture to spiritual conceptions."

[43]Pelikan, "'Spiritual Sense' of Scripture," 339.

[44]McGuckin, "Patterns of Biblical Exegesis," 46-47.

[45]Basil, *On the Holy Spirit* 7.16 (NPNF[2] 8:10). Italics added.

[46]Ibid., 2.4 (NPNF[2] 8:3).

[47]Pelikan, "'Spiritual Sense' of Scripture," 340. See Basil, *On the Holy Spirit* 6.15 (NPNF[2] 8:9). "What excuse can be found for their attack upon Scripture, shameless as their antagonism is."

Exodus 34 where Moses would place a veil over his face after speaking
with Yahweh on Sinai, Basil continues:

> But to leave no ground for objection, I will quote the actual words of the
> Apostle; – "For even unto this day remaineth the same veil untaken away
> in the reading of the Old Testament, which veil is done away in Christ.
> . . . Nevertheless, when it shall turn to the Lord, the veil shall be taken
> away. Now the Lord is that Spirit" [2 Cor 3:14-17]. Why does he speak thus?
> Because he who abides in the bare sense of the letter, and in it busies
> himself with the observances of the Law, has, as it were, got his own heart
> enveloped in the Jewish acceptance of the letter, like a veil; and this be-
> falls him because of his ignorance that the bodily observance of the Law
> is done away by the presence of Christ, in that for the future the types
> are transferred to the reality. Lamps are made needless by the advent of
> the sun; and, on the appearance of the truth, the occupation of the Law
> is gone, and prophecy is hushed into silence. He, on the contrary, who
> has been empowered to look down into the depth of the meaning of the
> Law, and, after passing through the obscurity of the letter, as through a
> veil, to arrive within things unspeakable, is like Moses taking off the veil
> when he spoke with God. He, too, turns from the letter to the Spirit. So
> with the veil on the face of Moses corresponds the obscurity of the
> teaching of the Law, and spiritual contemplation with the turning to the
> Lord. He, then, who in the reading of the Law takes away the letter and
> turns to the Lord, – and the Lord is now called the Spirit, – becomes
> moreover like Moses, who had his face glorified by the manifestation of
> God. For just as objects which lie near brilliant colours are themselves
> tinted by the brightness which is shed around, so is he who fixes his gaze
> firmly on the Spirit by the Spirit's glory somehow transfigured into
> greater splendour, having his heart lighted up, as it were, by some light
> streaming from the truth of the Spirit. And, this is "being changed from
> the glory" of the Spirit "into" His own "glory," not in niggard degree, nor
> dimly and indistinctly, but as we might expect any one to be who is en-
> lightened by the Spirit. Do you not, O man, fear the Apostle when he says
> "Ye are the temple of God, and the Spirit of God dwelleth in you" [1 Cor
> 3:16]? Could he ever have brooked to honour with the title of "temple"
> the quarters of a slave? How can he who calls Scripture "God-inspired"
> [2 Tim 3:16], because it was written through the inspiration of the Spirit,
> use the language of one who insults and belittles Him?[48]

---

[48]Basil, *On the Holy Spirit* 21.52 (NPNF[2] 8:33-34).

Does this Basil sound like one who would advocate an overly literal approach to Scripture and shun the spiritual/allegorical?

We also get a window into Basil's thinking about allegory from his Homily on Psalm 28 (LXX) (Homily 13), preached sometime during his episcopacy in Caesarea (370–379).[49] The title of the psalm is "A Psalm of David at the Finishing of the Tabernacle," and it is primarily about a charge delivered to the priests and Levites as they leave their sacred offices.[50] In the homily Basil asks what the psalm's title means:

> Let us consider what the finishing is and what the tabernacle is, in order that we may be able to meditate on the meaning of the psalm. Now, as regards the history, it will seem that the order was given to the priests and Levites who had acquitted themselves of the work to remember what they ought to prepare for the divine service. Scripture, furthermore, solemnly declares to those going out and departing from the tabernacle what it is proper for them to prepare and to have for their assembly on the following day: namely, "offspring of rams, glory and honor, glory to his name"; likewise it declares that nowhere else is it becoming to worship except in the court of the Lord and in the place of holiness. But, according to our mind which contemplates the sublime and makes the law familiar to us through a meaning which is noble and fitted to the divine Scripture, this occurs to us: the ram does *not* mean the male among the sheep; *nor* the tabernacle, the building constructed from this inanimate material; and the going out from the tabernacle does *not* mean the departure from the temple; but, the tabernacle for us is this body, as the Apostle taught us when he said: "We who are in this tabernacle sigh" [2 Cor 5:4]. And again, the psalm: "Nor shall the scourge come near thy dwelling" [Ps 90:10 LXX]. And the finishing of the tabernacle is the departure from this life, for which Scripture bids us to be prepared, bringing this thing and that to the Lord, since, indeed, our labor here is our provision for the future life. And that one who here bears glory and honor to the Lord through his good works will treasure up for himself glory and honor according to the just requital of the Judge.[51]

---

[49]Basil, Homily 13: *A Psalm of David at the Finishing of the Tabernacle,* in Saint Basil, *Exegetic Homilies,* trans. Agnes Clare Way, FC 46 (Washington, DC: Catholic University of America Press, 1963), 193-211.

[50]The summary here is dependent on Jackson, "Introduction" (NPNF² 8:xlv).

[51]Basil, Homily 13 (FC 46:193-94). Italics added.

It is reasonable to ask, Would one who advocates a strictly literal interpretation of Scripture and shuns a figural or allegorical reading make the explicit claim that the meaning of the passage is not what is written on the page? Yet for Basil, "The ram does *not* mean the male among the sheep; *nor* the tabernacle, the building constructed from this inanimate material; and the going out from the tabernacle does *not* mean the departure from the temple."

There are other non-Hexaemeral passages that could be offered wherein Basil advocates the rejection of a wooden literal interpretation in favor of a figural or allegorical reading.[52] Often these occur with extended justification for why a figural reading is necessitated because of the inspiration of sacred Scripture.[53]

LITERAL AND ALLEGORICAL IN BASIL'S *HEXAEMERON*

We do not necessarily see the abovementioned type of spiritual reading of Scripture in Basil's homilies on the Hexaemeron. But there is good reason to reconsider Mook's attempt at attributing a synonymous relationship between Basil's understanding of "literal" or "common usage" and ours (or, better said, those who claim Basil is "on their side" in a literal rendering of Genesis). There are, in other words, places where Basil goes beyond the bare letter of the words and encourages his hearers to do the same. These basically fall into three types: First, the comparison of natural phenomena with human situations for the purpose of moral lessons;[54] second, calls to a deeper "mystical" reading; and third, explicit spiritual readings or allegorizations.

---

[52]See, for example, Letter 8.6-7 (NPNF[2] 8:118-19), which contains an elaborate allegorization of Acts 1:7; and Homily 14 on Psalm 29, which, according to Hughes Oliphant Old, "is to a large extent an example of how allegory was used in the ancient Church to interpret Scripture." See Old, *Reading and Preaching of the Scriptures*, 2:41.

[53]A clear example of this was seen in *On the Holy Spirit*.

[54]Bouteneff, *Beginnings*, 136.

THE COMPARISON OF NATURAL PHENOMENA
WITH HUMAN SITUATIONS

The first type of nonliteral reading, the comparison of natural phenomena with human situations for the purpose of moral lessons, occurs most often in the *Hexaemeron*. These interpretations do not really begin in earnest until Homily 5, which corresponds to Genesis 1:11 with the creation of vegetation and, especially, animals. Typically, Basil starts with a verse that indicates the creation of a natural phenomenon, plant, or animal; details some specifics that would be included in that creation; and then compares some specific traits that humans should avoid or exemplify. Examples of this are plentiful in the *Hexaemeron*, so the following are selective.[55]

In Homily 6 Basil comments on Genesis 1:16, "And God made two great lights."[56] He uses the verse as a springboard to speak about the moon and stars. When he turns specifically to the moon, he notices that although it is immense and bright, its size does not always remain visible because of shadows.[57] He explains that there is a "hidden reason" for the varied shape of the moon: "In truth, it is so as to provide for us a clear example of our nature. For, nothing human is stable."[58] Thus, the moon actually teaches us about our own vicissitudes, and because of that, we should be wary of pride in our successes, power, and wealth. In the flesh there is constant change, so one should care for the soul, "whose good is unchangeable."[59] The moon is then compared to the human soul. One who is troubled that the moon diminishes its splendor in its waxing and waning should be distressed much more so by the soul, which "having possessed virtue, through neglect destroys its beauty and never remains in the same disposition, but turns and changes constantly through fickleness of mind."[60]

---

[55]Other examples include *Hexaemeron* 5.2, 6; 6.1, 10; 7.3-5; 8.1, 4-5; 9.2-3, 11, 15.

[56]Basil, *Hexaemeron* 6.9 (FC 46:97).

[57]Ibid., 6.10 (FC 46:99).

[58]Ibid. (FC 46:100).

[59]Ibid.

[60]Ibid.

In Homily 7 Basil opens by quoting Genesis 1:20: "Then God said, 'Let the waters bring forth crawling creatures' of different kinds 'that have life, and winged creatures' of different kinds 'that fly below the firmament of the heavens.'" As Basil details some of the creatures involved here, he begins to focus on fish. The majority of fish, he explains, eat one another. Smaller fish become food for larger fish. Sometimes after one fish eats a smaller one, the first fish then becomes prey of an even larger fish. Thus, both fish are carried away in the stomach of the last. From this natural phenomenon, Basil draws a moral lesson:

> Now what else do we men do in the oppression of our inferiors? How does he differ from the last fish, who with a greedy love of riches swallows up the weak in the folds of his insatiable avarice? That man held the possessions of the poor man; you seizing him, made him a part of your abundance. You have clearly shown yourself more unjust than the unjust man and more grasping than the greedy man. Beware, lest the same end as that of the fish awaits you – somewhere a fishhook, a snare, or a net. Surely, if we have committed many unjust deeds, we shall not escape the final retribution.[61]

But Basil does not end there. He continues to compare evildoers to the crab and exhorts his hearers to avoid imitating it as he explains how the crab preys on the oyster. "Nature" has protected the tender flesh of the oyster with a hard shell that is very difficult to penetrate because its closed shell envelopes together. The crab waits until it sees the oyster warming itself in the sun with its shell open. Then it stealthily inserts a small pebble in the hinge of the oyster shell, which prevents it from closing its shell. In doing so the crab finds its way into the shell by inventiveness rather than strength. Basil concludes:

> Now, I want you, although emulating the crabs' acquisitiveness and their inventiveness, to abstain from injury to your neighbors. He who approaches his brother with deceit, who adds to the troubles of his neighbors, and who delights in others' misfortunes, is like the crab. Avoid

---

[61]Ibid., 7.3 (FC 46:109).

the imitation of those who by their conduct convict themselves. Poverty with an honest sufficiency is preferred by the wise to all pleasure.[62]

Examples of this type could be multiplied, even in this section of Homily 7 where Basil exhorts his hearers not to be like the octopus. But from the examples provided one gets a clear sense of how Basil compares natural phenomena with human situations for the purpose of moral lessons. This is the first way Basil goes beyond the common usage or bare letter of the words in the text.[63] The question that must be asked here is, what kind of reading of Scripture is this – literal or allegorical? Could the author of "Genesis Means What It Says" affirm that these examples are evidence that Basil believed the "words are to be understood by their plain meaning"? Could James Mook evaluate Basil's reading here and claim that it is "advocating instead a humble acceptance of the 'common sense,' the 'literal sense' of Scripture 'as it has been written'"?[64] If, as Mook claims, there was a sharp division in the early church between the literalists and the allegorists, on what side would Basil fall in light of this type of reading?[65] Perhaps Mook's "common reading" is not what Basil would consider the "common reading"; if so, what determines how Mook conceives of this, and what controls how Basil understands it?

CALLS TO A DEEPER MEANING

The second type of passage where Basil goes beyond the bare letter occurs through general calls for his hearers to move on to a deeper reading of Scripture.

---

[62]Ibid. (FC 46:110).

[63]Paul Blowers states that these are all "reminiscent of the moral and spiritual lessons drawn from nature in the *Physiologus*" (*Drama of the Divine Economy: Creator and Creation in Early Christian Theology and Piety*, Oxford Early Christian Studies [Oxford: Oxford University Press], 128). The *Physiologus* was a second-century CE anthology that uses traits from animals to draw moral or theological lessons from their behavior or illustrate texts from Scripture. See Robert M. Grant, *Early Christians and Animals* (London: Routledge, 1999), 44-72.

[64]Mook, "Church Fathers on Genesis," 30. Mook employs the translation by Jackson found in NPNF[2] vol. 8.

[65]Mook, "Church Fathers on Genesis," 29.

In Homily 3.2 Basil quotes Genesis 1:6, "Then God said, 'Let there be a firmament in the midst of the waters to divide the waters.'" He stresses the importance of understanding how God speaks, and refers to his previous homily the day before when he preached on the words of God, "Let there be light." Does God, Basil asks, speak in our manner?

> Is the image of the objects first formed in his intellect, then, after they have been pictured in his mind, does he make them known by selecting from substances the distinguishing marks characteristic of each? Finally, handing over the concepts to the vocal organs for their service, does he thus manifest His hidden thought by striking the air with the articulate movement of his voice?[66]

It is "fantastic" to say that God needs such a "roundabout way" to express his thoughts.

Instead, true religion insists that this is a reference to the Word of God. The passage, according to Basil, shows that God not only wanted creation accomplished but also brought it forth through a coworker. Scripture could have simply related everything as it began – the heavens and the earth, light, firmament. But by indicating God as speaking and commanding, "it indicates silently Him to whom He gives the command and to whom He speaks, not because it begrudges us the knowledge, but that it may inflame us to a desire by the very means by which it suggests some traces and indications of the mystery."[67] He continues with the explicit claim that in these passages Scripture leads us, in an orderly way, to the "idea of the Only-begotten."[68] "This way of speaking has been wisely and skillfully employed so as to rouse our mind to an inquiry of the Person to whom in these words are directed."[69] Apparently for Basil, a reading according to common use includes searching for "indications of the mystery." In this case, the mystery is the agency of Christ in the creation.

---

[66]Basil, *Hexaemeron* 3.2 (FC 46:38).
[67]Ibid. (FC 46:39).
[68]Ibid.
[69]Ibid.

In Homily 6.2 Basil quotes Genesis 1:14-15, "And God said, 'Let there be lights in the firmament of the heavens for the illumination of the earth, to separate the day from night.'" He points out that neither the sun nor the moon were in existence yet. This was so that people ignorant of God might not claim the sun as the first cause, regarding it as the producer of all that grows. This was the reason why God, on the fourth day said, "'Let there be light' . . . and God made two lights." In keeping with what Basil claimed about the Logos as God's "coworker" in creation, Basil asks his hearers to consider who it was who spoke and made: "Do you not notice in these words the double Person? Everywhere in history the teachings of theology are mystically interspersed."[70]

At the end of the same homily Basil wraps up his discussion of the luminaries indicated in Genesis 1:16. As we have seen above, in Homily 6 Basil goes beyond the bare meaning of the words in ways that correspond to our first two types of examples. He is convinced that "none of the divinely inspired words, even as much as a syllable, is an idle word."[71] Thus, God's creation is meant to lead us on to him.

> May He who has granted us intelligence to learn of the great wisdom of the Artificer from the most insignificant objects of creation permit us to receive loftier concepts of the Creator from the mighty objects of creation. And yet, in comparison with the Creator, the sun and moon possess the reason of a gnat or an ant. Truly, it is not possible to attain a worthy view of the God of the universe from these things, but to be led on by them, as also by each of the tiniest plants and animals to some slight and faint impression of Him.[72]

Genesis 1:20, "Let the waters bring forth crawling creatures of different kinds that have life and winged creatures that fly above the earth under the firmament of the heavens," gives Basil occasion to speak

---

[70]Ibid., 6.2 (FC 46:85).
[71]Ibid., 6.11 (FC 46:101).
[72]Ibid. (FC 46:102-3). Another example comes in *Hexaemeron* 4.6 (FC 46:63-65), where the sea is compared to the church.

about birds in Homily 8.[73] Amid the detailed delineation of bird species he has a short section on how some birds produce offspring. Some birds, he explains, do not need union with males for conception, but others that produce eggs without copulation are sterile. He notes the specific example of the vulture as a bird that hatches without coition. Then, as a summary of what he has just stated, he draws a "special observation from the history of the birds":

> If you ever see any persons laughing at our mystery, as though it were impossible and contrary to nature for a virgin to give birth while her virginity itself was preserved immaculate, you may consider that God, who is pleased to save the faithful by the foolishness of our preaching [1 Cor 1:21], first set forth innumerable reasons from nature for our beliefs in His wonders.[74]

In the same homily, we see something similar, except this time, instead of focusing on the virgin birth, he focuses on the resurrection. "The words of Scripture, if simply read, are a few short syllables: 'Let the waters bring forth winged creatures that fly above the earth under the firmament of the heavens'; but when the meaning in the words is explained, then the great marvel of the wisdom of the Creator appears."[75] This call to go deeper leads him to cite Genesis 1:24 and speak about certain land animals.[76] He expresses surprise that so many distrust Paul concerning the transformation made at the resurrection when they have observed creatures of the air changing their forms.[77] He then speaks about the Indian silkworm, which changes into a caterpillar and then a flying insect with "light, wide, metallic wings." So, Basil continues, when people unwind the products of these animals, "recall the metamorphoses in this creature, conceive a clear idea of the resurrection, and do

---

[73]Ibid., 8.2 (FC 46:120).
[74]Ibid., 8.6 (FC 46:128).
[75]Ibid., 8.8 (FC 46:132).
[76]Gen 1:24: "Let the earth bring forth living creatures, cattle and wild animals and crawling creatures of different kinds."
[77]Col 3:4.

not refuse to believe the change which Paul announces for all men."[78]
Again, the question must be asked what kind of reading of Scripture this
is—literal or allegorical?

EXPLICIT SPIRITUALIZATION/ALLEGORIZATION

In the third type of passage where Basil goes beyond the bare letter, he
uses explicit spiritualization or allegorization.

In Homily 5 Basil directs his attention to the germination of the
earth. Genesis 1:11 gives him the occasion to comment on trees, especially ones that produce fruit.[79] These trees, like the vine that produces
wine or the olive tree that can give us olive oil, contribute in a significant way to our lives. This leads Basil to comment a bit more on the
vine, which, he says, should remind us of our nature. Immediately
coming to his mind in this respect is the passage in the Gospel of John
about the vine and the branches.[80] Fruit is expected of the branches
(us), otherwise they will be cut off and thrown into the fire.

The reference to John 15 opens up whole new vistas as Basil criss-
crosses over several other scriptural passages intended to show what
he believes is anchored in the meaning of Genesis 1:11. He states that
in Scripture, the souls of humanity are often compared to vineyards.[81]
Even though none of the passages cited by Basil thus far connects the
vineyard to human souls, he makes that connection: "Evidently, He [the
Lord] calls the human souls the vineyard, about which He has put as a

---

[78]Basil, *Hexaemeron* 8.8 (FC 46:132).

[79]Basil, *Hexaemeron* 5.6 (FC 46:74), "And the fruit tree that bears fruit containing seed of its own kind and of its own likeness on the earth."

[80]Basil neither quotes the passage in full nor gives the reference, but it is quite clear he is referring to Jn 15:1-6: "I am the true vine, and my Father is the vinedresser. Every branch of mine
that bears no fruit, he takes away, and every branch that does bear fruit he prunes, that it
may bear more fruit. You are already made clean by the word which I have spoken to you.
Abide in me, and I in you. As the branch cannot bear fruit by itself, unless it abides in the
vine, neither can you, unless you abide in me. I am the vine, you are the branches. He who
abides in me, and I in him, he it is that bears much fruit, for apart from me you can do nothing. If a man does not abide in me, he is cast forth as a branch and withers; and the branches
are gathered, thrown into the fire and burned."

[81]In this regard he cites Is 5:1: "My beloved had a vineyard on a very fertile hill." He also cites
what appears to be Mt 21:33: "I planted a vineyard, and put a hedge about it."

hedge the security arising from his commandments and the custody of His angels."[82] But Basil is still not finished teasing out the meaning as he moves on to speak about the apostles, prophets, and teachers as "props" who were "planted for us" in the establishment of the church.[83] These "blessed men of old" are examples meant to lead our thoughts upward because the Lord did not allow them to be tossed on the ground and trampled. We are exhorted to cling to our "neighbors" with love, "like the tendrils of the vine, and to rest upon them, so that, keeping our desires always heavenward, we may, like certain climbing vines, reach the upmost heights of the loftiest teachings."[84]

But Basil is still not finished wringing meaning out of Genesis 1:11 based on its suggestion of the vine. In an allusion to Luke 13:8, he claims that the Lord asks us to "permit ourselves to be dug about."[85] To be "dug about," for Basil, means putting aside the burdensome cares of the world. The one who has laid aside the love and desire of possessions and pride has been "dug about." This all culminates in Basil's final words on this subject, which bring it back around to the vine in Genesis 1:11, his original starting point:

> We must not, according to the meaning of the proverb, run to wood, that is, live our lives ostentatiously, or eagerly seek praise from those outside, but we must be fruitful, preserving for the true Farmer the proof of our works. But, you also, be "as a fruitful olive tree in the house of God," [Ps 52:8] never destitute of hope, but always having about you the rich assurance of safety through faith. Thus, indeed, you will imitate the eternal verdure of this plant and emulate its fruitfulness, bestowing bounteous alms on every occasion.[86]

The insistence that Basil was a literalist is also challenged in the

---

[82]Basil, *Hexaemeron* 5.6 (FC 46:76). To further drive this point, he cites Ps 34:7: "The angel of the Lord shall encamp round about them that fear him."

[83]1 Cor 12:28.

[84]Basil, *Hexaemeron* 5.6 (FC 46:76).

[85]Lk 13:8: "And he answered him, 'Let it alone, sir, this year also, till I dig about it and put on manure.'"

[86]Basil, *Hexaemeron* 5.6 (FC 46:76-77).

aforementioned Homily 6, which deals explicitly with the creation of the lights in the heavens. After explaining how Genesis 1:14-15 points to the Son as the Father's agent in creation, Basil comments on why the creation of luminaries is necessary.[87] In the Septuagint translation used by Basil, Genesis 1:14 states that the creation of luminaries is necessary "for the illumination of the earth." He recognizes that light had already been created and asks why the text now says that the sun was created to give light. He acknowledges the peculiarity here and even asks that his hearers refrain from laughing, "if indeed we do not follow the choice of words nor pursue rhythm in the arrangement of them."[88] Basil notes that the text uses the word *illumination* (*phausis*) here, not *light* (*phōtismos*). Because of this there is no conflict with what Genesis earlier stated about "light" (*phōs*). The actual nature of light was created in Genesis 1:3, but here, in Genesis 1:14, the sun was made to be a vehicle for that first created light. Fire has the power to give light, but the lamp gives light to those who need it. Thus, in the case of Genesis 1:14, the luminaries were created as a vehicle for the pure, clear, immaterial light.

Thus far, there is nothing here that would necessarily qualify as spiritual or allegorical. But immediately following this clarification on how the sun can be created after light, Basil takes a deeper, spiritual track:

> And, just as the Apostle says that there are certain lights in the world [see Phil 2:15], but the true Light of the world is something else, and by participation in it holy men become the lights of the souls whom they have taught, drawing them out from the darkness of ignorance, so also now, having prepared this sun for that most bright light, the Creator of the universe has lighted it around the world.[89]

In what sense can either of these two examples from the *Hexaemeron* be called a "literal" reading?[90] Is Basil's explanation of the vine, based

---

[87]Ibid., 6.2 (FC 46:85-86).
[88]Ibid.
[89]Ibid.
[90]Another extended example could be mined from Homily 2.8, which we will examine in detail in chapter six.

on the trees in Genesis 1:11, a reading that is accurately described as plain? Does Basil's move from the creation of light, and then the sun, to participation in the true Light of the world lead us to agree with James R. Mook that Basil advocated and exemplified a humble acceptance of the "common sense" of Genesis?[91] Again, what kind of reading of Scripture is this – literal or allegorical?

THE PASSAGES USED AS PROOF TEXTS

The beginning of this chapter cited a phrase from Homily 9 in Basil's *Hexaemeron* that appears to reject allegory. As already indicated, there are actually two passages in his nine homilies that explicitly mention allegory. We will deal with these in the order they appear in the *Hexaemeron*.

Homily 3.9.

We have also some argument concerning the division of the waters with those writers of the Church who, on a pretext of the spiritual sense [*profasei anagoges*] and of more sublime concepts, have recourse to allegories, saying that spiritual and incorporeal powers are signified figuratively by the waters, that the more excellent have remained up above the firmament, but the malignant remain below in the terrestrial and material regions. For this reason, they say, the waters above the heavens praise God; that is, the good powers, being worthy because of the purity of their reasoning, pay to the Creator becoming praise. But, the waters under the heavens are the spirits of malice, which have fallen down from their natural height to the depth of wickedness. Inasmuch as these are tumultuous and factious and agitated by the uproar of the passions, they are named "sea" from the instability and inconstancy of their voluntary movements. Dismissing such explanations as dream interpretations and old women's tales, let us consider water as water, and let us receive the separation that was made beneath the firmament according to the reason given us.[92]

Immediately preceding this section Basil compares "the simplicity and lack of artifice of spiritual discourse with the futile questioning

---

[91]Mook, "Church Fathers on Genesis," 30.
[92]Basil, *Hexaemeron* 3.9 (FC 46:51-52).

of philosophers about the heavens."[93] The philosophers and scientists offer a "forced persuasiveness," while Basil offers truth that is "set forth bare of artifices." Because of this he turns his attention to church writers. The problem is that some of these writers "have recourse to allegories."

Some scholars argue that Basil actually has Origen in mind here as the writer in the church who sought refuge in allegory.[94] This can perhaps be attributed to the claims of two important church fathers — Jerome (347–420) and Epiphanius (ca. 315–403). In Epiphanius's *Panarion*[95] and Jerome's *To Pammachius Against John of Jerusalem*,[96] Origen is taken to task for allegorizing paradise and the waters of Genesis 1.[97] Jerome and Epiphanius's protest of Origen's allegorizing of the heavenly waters and the waters below the earth, combined with the claims of some scholars that Basil is here indicating (and condemning) Origen, had a strong influence.

But the conclusion that Basil has Origen specifically in mind in Homily 3 has not gone unchallenged. Richard Lim has pointed out that there are parts of Basil's criticism of allegory that simply could not be

---

[93]Ibid., 3.8 (FC 46:51).

[94]What follows draws from Lim, "Politics of Interpretation," 355-57. Lim cites Jean Pépin as a scholar who makes this assumption. See *Mythe et Allegorié: Les origenes grecques et les contestations judé-chrétiennes* (Paris: Études Augustiniennes, 1958), 453.

[95]Epiphanius of Salamis, *The Panarion of Epiphanius of Salamis: Books II and III (Sects 47–80, De Fide)*, Nag Hammadi and Manichaean Studies, trans. Frank Williams (Leiden: Brill, 1994). "Finally, [Origen] interprets whatever he can allegorically — Paradise, its waters, the waters above the heavens, the water under the earth. He never stops saying these ridiculous things and others like them" (64.4.11, p. 136).

[96]Jerome, *To Pammachius Against John of Jerusalem* 7 (NPNF² 6:428). "Sixthly, he so allegorises Paradise as to destroy historical truth, understanding angels instead of trees, heavenly virtues instead of rivers, and he overthrows all that is contained in the history of Paradise by his figurative interpretation. Seventhly, he thinks that the waters which are said in Scripture to be above the heavens are holy and supernal essences, while those which are above the earth and beneath the earth are, on the contrary, demoniacal essences."

[97]See also Jerome's translation of a letter, *From Epiphanius, Bishop of Salamis, in Cyprus, to John, Bishop of Jerusalem* (NPNF² 6:86-87): "Must not every one reject and despise such special pleading as that by which Origen says of the waters that are above the firmament that they are not waters, but heroic beings of angelic power, and again of the waters that are over the earth — that is, below the firmament — that they are potencies of the contrary sort — that is, demons?"

applied to Origen.[98] For example, in Origen's *Homilies on Genesis*, after citing Genesis 1:7-8, he writes the following:[99]

> Let each of you, therefore, be zealous to become a divider of that water which is above and that which is below. The purpose, of course, is that, attaining an understanding and participation in that spiritual water which is above the firmament one may draw forth "from within himself rivers of living water springing up into life eternal," [cf. Jn 4:14; 7:38] removed without doubt and separated from that water which is below, that is, the water of the abyss in which darkness is said to be, in which "the prince of this world" and the adversary, "the dragon and his angels" dwell [cf. Mt 25:41; Jn 12:31; 1 Pet 5:8; Rev 12:7-9; 20:2-3], as was indicated above.[100]

Here Origen is employing a common and well-accepted patristic way of reading Scripture.[101] He comes to his conclusion by connecting Revelation 20:3 with other Scriptures.[102] Is this way of reading really any different than that employed by Basil himself in Homily 5.6? Can Origen really be accused of "taking refuge in allegory" and being "ashamed of the scriptures" when he seeks to "draw intertextually on other parts of scriptures to throw light on the particular line in Genesis"?[103] Would Basil accuse Origen of this when he himself employs it in his own *Hexaemeron*?

Even a broader look at Origen's exegetical discussions in this *Homily on Genesis* cannot support the accusation that Basil has Origen specifically in mind.[104] There is no indication in Origen, for example, that he equates the "waters under the heavens" with "the spirits of malice,

---

[98]Lim, "Politics of Interpretation," 355.

[99]Gen 1:7-8: "So God made the dome and separated the waters that were under the dome from the waters that were above the dome. And it was so. God called the dome Sky. And there was evening and there was morning, the second day."

[100]Origen, *Homily* 1.2 (FC 71:50). All quotations of *Homilies on Genesis* are from Origen, *Homilies on Genesis and Exodus*, trans. Ronald E. Heine, FC 71 (Washington, DC: Catholic University of America Press, 2010).

[101]See Peter J. Leithart, *Deep Exegesis: The Mystery of Reading Scripture* (Waco, TX: Baylor University Press, 2009).

[102]Rev 20:3: "And threw him into the pit, and locked and sealed it over him, so that he would deceive the nations no more, until the thousand years were ended. After that he must be let out for a little while."

[103]Lim, "Politics of Interpretation," 356.

[104]Ibid.

which have fallen down from their natural height to the depth of wickedness."[105] Thus, Lim concludes that Basil does not, in fact, have Origen particularly in mind; rather, he refers to "certain later allegorists who might, or might not, have been specifically elaborating Origen's exegesis."[106]

To support his conclusion, Lim points to the actual concern for Basil's reservation toward allegory in this sense. If we read a bit further in the passage (*Hexaemeron* 3.9) we see that Basil reveals why the allegorical interpretations he mentions are not satisfactory. The allegorization to which he objects is of the water under the heaven meaning "spirits of malice." These spirits who fell are called "sea" because of their unstable and inconsistent movements. It is these kinds of meanings that are "dream interpretations and old women's tales."[107] It was popular in Neoplatonic circles to associate demons with the movements of the sea.[108] Thus we see that Origen does not, in fact, allegorize in the way indicated by Basil. However, certain Neoplatonists do, and it may be that this kind of reading is creeping in to "those writers in the Church."[109]

Lim also argues that even though Basil rejects the allegorical interpretation of the above passage, we should note that he does not offer a more "literal" interpretation in its place.[110] The allegorization is dismissed as a dream or fable, not worthy of consideration. He does the same thing a bit later in the same homily.[111] But even then he gives a nod to the deeper spiritual meaning of Scripture and concludes the

---

[105]Basil, *Hexaemeron* 3.9 (FC 46:52).

[106]Lim, "Politics of Interpretation," 356.

[107]Basil, *Hexaemeron* 3.9 (FC 46:52).

[108]Lim, "Politics of Interpretation," 356. Lim states, "Proclus, in his rhetorical defense of Homer against Socrates' charges in the *Republic*, describes the demons as connected with the 'disorderly and erratic movement' in the material world."

[109]Basil, *Hexaemeron* 3.9 (FC 46:51).

[110]Lim, "Politics of Interpretation," 365.

[111]Basil, *Hexaemeron* 3.9 (FC 46:52). "And if someone says that the heavens are speculative powers, and the firmament, active powers productive of the good, we accept the expression as neatly said, but we will not concede that it is altogether true."

section, which began with a criticism of certain kinds of allegory, in
the following way:

> The meaning in these words, however, accepted by speculative minds, is
> a fulfillment of the praise of the Creator. Not only the water which is
> above the heavens, as if holding the first place in honor because of the
> pre-eminence added to it from its excellence, fulfills the praise of God,
> but, "Praise him," the psalmist says [Ps 148:7], "from the earth, ye dragons,
> and all ye deeps." So that even the deep, which those who speak alle-
> gories relegated to the inferior portion, was not itself judged deserving
> of rejection by the psalmist, since it was admitted to the general chorus
> of creation; but even it harmoniously sings a hymn of praise to the
> Creator through the language assigned to it.

Hence it is very likely that Origen is not in Basil's mind here in this
first apparent rejection of allegory. Further, it is difficult to see it as a
wholesale rejection of allegorical reading. Given what we have already
learned about Basil's approach to Scripture with regard to allegorical
or spiritual reading, a wholesale rejection of allegory would certainly
be out of place in his own thinking.

Rather than reading *Hexaemeron* 3.9 as a rejection of Origen's allegorization of Scripture, we actually begin to see Basil's strong connection
to him. It is very likely that Basil did not have Origen in mind here as
a church writer prone to unacceptable allegories. But we also get a
strong sense that Basil employed some of the same reading strategies
as Origen. When we turn to Basil's second apparent rejection of allegory, that strong connection between Origen's reading strategy and
Basil's is confirmed.

Homily 9.1.

I know the laws of allegory, although I did not invent them of myself, but
have met them in the works of others. Those who do not admit the
common meaning of the Scriptures say that water is not water, but some
other nature, and they explain a plant and a fish according to their
opinion. They describe also the production of reptiles and wild animals,
changing it according to their own notions, just like the dream interpreters, who interpret for their own ends the appearances in their

dreams. When I hear "grass," I think of grass, and in the same manner I understand everything as it is said, a plant, a fish, a wild animal, and an ox. "Indeed, I am not ashamed of the gospel." . . . This is the thing which they seem to me to have been unaware, who have attempted by false arguments and allegorical interpretations to bestow on Scripture a dignity of their own imagining. But, theirs is the attitude of one who considers himself wiser than the revelations of the Spirit and introduces his own ideas in pretense of an explanation. Therefore, let it be understood as it has been written.[112]

This passage could be called the *locus classicus* for those who employ Basil as a literalist. The problem is that no reputable scholar of the church fathers would make the argument that Basil was a literalist. In fact, the passage has received attention because it does not appear to fit with what we know about Basil and his exegetical approach.

The passage cited above is preceded by an interesting paragraph that opens the homily. In the opening, Basil compares himself to Elisha, who served his friends wild plants.[113] Some believe that this is a response to criticism he received about his comments on allegory in Homily 3.9.[114] Just as the wild plants Elisha served was poor fare, Basil pictures his congregation as failing to appreciate his more literal reading given only a few days earlier.[115] He counsels that they not ignore it: "Elisha was by no means rejected as a poor host by his contemporaries in spite of the fact that he feasted his friends on wild plants."[116] This apparent pushback from Basil's congregation *against* a more literal interpretation may suggest that the allegorical method was the commonly accepted equipment of a church leader of Basil's stature.[117]

---

[112]Ibid., 9.1 (FC 46:135).

[113]Ibid. The reference is probably to 2 Kings 4:39: "One of them went out into the field to gather herbs; he found a wild vine and gathered from it a lapful of wild gourds, and came and cut them up into the pot of stew, not knowing what they were."

[114]Bouteneff, *Beginnings*, 129; Lim, "Politics of Interpretation," 357; and McGuckin, "Patterns of Biblical Exegesis," 45.

[115]The significance of this point should not escape us. That Basil may be criticized here for a literal interpretation begs the question about what exactly "literal" is in this context.

[116]Basil, *Hexaemeron* 9.1 (FC 46:135).

[117]Lim, "Politics of Interpretation," 357.

Basil cannot simply be categorized as either an allegorist or literalist.[118] All agree that Basil does adopt a more literal approach in these homilies. But, as we saw in the previous chapter, we cannot simply equate our understanding of literal with Basil's. The real issue has been the question of why he tends toward a literal reading.[119] Just as it did in connection with the apparent rejection of allegory in *Hexaemeron* 3.9, Origen's name comes up here as Basil's referent in his words against allegory in *Hexaemeron* 9.1. Further, just as in our discussion of *Hexaemeron* 3.9, the connection is important because if Basil did *not* have Origen in mind in these remarks, and if he actually accepted Origen's basic scheme of how to read Scripture, this would significantly curtail any wholesale rejection of allegory. In other words, it would suggest, perhaps, that there are acceptable and unacceptable allegorical readings for Basil.

Basil's connection to Origen is well known. An indirect connection comes from his grandmother's conversion to Christianity as a result of the influence of Origen's most devoted follower and propagator of his thought, Gregory Thaumaturgus (the "Wonder Worker").[120] Basil states, "What clearer evidence can there be of my faith, than that I was brought up by my grandmother, blessed woman, who came from you? I mean the celebrated Macrina who taught me the words of the blessed Gregory; which, as far as memory had preserved down to her day, she cherished herself, while she fashioned and formed me, while yet a child, upon the doctrines of piety." This suggests that Basil learned Origen through Thaumaturgus.

---

[118]Sandwell, "How to Teach Genesis 1.1-19," 543. See Frances M. Young, "The Rhetorical Schools and Their Influence on Patristic Exegesis," in *The Making of Orthodoxy: Essays in Honour of Henry Chadwick*, ed. Henry Chadwick and Rowan Williams (Cambridge: Cambridge University Press, 1989), 195-96; and Frances M. Young, *Biblical Exegesis and the Formation of Christian Culture* (Peabody, MA: Hendrickson, 2002), 186.

[119]I should add that it is also important to keep in mind what *literal* means in Basil's context. See chapter three.

[120]Basil, *Letter* 204.6 (NPNF² 8:245). See also Basil's very complimentary words about Gregory in *On the Holy Spirit* 29.74 (NPNF² 8:47).

Explicit references to Origen in Basil's writings are rare, although we should note that it is common among the Fathers not to refer to one another by name. He refers explicitly to Origen only once.[121] This, of course, should not rule out the probability that Basil knew and used Origen's work.[122]

Another direct connection is found in Basil and Gregory of Nazianzus's compilation of extracts from Origen's writings called the *Philocalia*.[123] The collection was published around 358 and contains, among others, noncontroversial texts on Origen's understanding of the inspiration and interpretation of Scripture.[124] Since a significant component of the *Philocalia* is devoted to Origen's exegetical principles, the divine inspiration of Scripture, and the importance of spiritual exegesis, this may suggest that Basil was not critical of it in 358, long before the preaching of his homilies on the six days of creation.[125]

A final and very significant connection between Origen and Basil is in the threefold division of scriptural interpretation that is explicitly expressed in the *Philocalia*. Its connection to Basil is made clear by Gregory of Nazianzus in his famous *Funeral Oration for Basil*:

> I will only say this of him. Whenever I handle his *Hexaemeron*, and take its words on my lips, I am brought into the presence of the Creator, and understand the words of creation, and admire the Creator more than before, using my teacher as my only means of sight. Whenever I take up his polemical works, I see the fire of Sodom, by which the wicked and rebellious tongues are reduced to ashes, or the tower of [Babel], impiously

---

[121]Basil, *On the Holy Spirit* 29.73 (NPNF[2] 8:46).

[122]McGuckin states: "Basil clearly knew Origen's work and often uses it quietly and unostentatiously" ("Patterns of Biblical Exegesis," 44).

[123]McGuckin (*Saint Gregory of Nazianzus: An Intellectual Biography* [Crestwood, NY: St. Vladimir's Seminary Press, 2001], 102-4; and McGuckin, "Patterns of Biblical Exegesis," 44-45) suggests that the work was largely Gregory's, not Basil's. This, however, does not negate the fact that Basil knew and approved of the work.

[124]The Greek text can be found in J. A. Robinson, *The Philocalia of Origen* (Cambridge: Cambridge University Press, 1893). English translation can be found in George Lewis, *The Philocalia of Origen: A Compilation of Selected Passages from Origen's Works Made by St. Gregory of Nazianzus and St. Basil of Caesarea* (Edinburgh: T&T Clark, 1911).

[125]Lim, "Politics of Interpretation," 351.

built, and righteously destroyed. Whenever I read his writings on the
Spirit, I find the God Whom I possess, and grow bold in my utterance of
the truth, from the support of his theology and contemplation. His other
treatises, in which he gives explanations for those who are shortsighted,
by a threefold inscription on the solid tablets of his heart, lead me on
from a mere literal or symbolical interpretation to a still wider view, as
I proceed from one depth to another, calling upon deep after deep, and
finding light after light, until I attain the highest pinnacle. When I study
his panegyrics on our athletes, I despise the body, and enjoy the society
of those whom he is praising, and rouse myself to the struggle. His moral
and practical discourses purify soul and body, making me a temple fit for
God, and an instrument struck by the Spirit, to celebrate by its strains
the glory and power of God. In fact, he reduces me to harmony and order,
and changes me by a Divine transformation.[126]

This is a reference to the well-established and accepted scheme first
laid out by Origen in *On First Principles* – a passage also included in the
Philocalia.[127]

In the *Philocalia*, Origen bemoans the "many mistakes" made by Jews
and heretics in reading Scripture.[128] The reason why they have made
many mistakes "appears to be that the Scripture on the spiritual side
is not understood, but is taken in the bare literal sense."[129] In an explanation not unlike Basil's appeal with respect to the Holy Spirit, Origen
claims that because the Scriptures are inspired by the Holy Spirit, we
must look for the spiritual meaning in them. He claims that the "right
way" to read Scriptures is anchored in Proverbs 22:20-21.[130] Just as a
human being consists of body, soul, and spirit, so too does Scripture,
which is given by God for the salvation of humankind. Origen states:

---

[126]Gregory of Nazianzus, *Oration* 43.67 ("Funeral Oration on the Great S. Basil, Bishop of Cæsarea
in Cappadocia") (NPNF² 7:417-18).

[127]Origen, *On First Principles* 4.2.4 (Butterworth, 274-77). All quotations from *On First Principles* are
from Origen, *On First Principles*, trans. G. W. Butterworth (Gloucester, MA: Peter Smith, 1973).

[128]*Philocalia* 1.8 (Lewis, 9-10).

[129]Ibid., 1.9 (Lewis, 10).

[130]Ibid., 1.11 (Lewis, 12). Prov 22:20-21 (LXX): "Now then, copy them for yourself three times over,
for counsel and knowledge on the surface of your heart. Therefore I teach you a true word
and good knowledge to heed in order that you may answer the words of truth to them who
question you."

"A man ought then in three ways to record in his own soul the purposes of the Holy Scriptures; that the simple may be edified by, as it were, the *flesh* of Scripture (for thus we designate the primary sense), the more advanced by its *soul*, and the perfect by the spiritual law, which has a shadow of the good things to come."[131]

Origen uses an illustration from the then highly esteemed Shepherd of Hermas to explain the division.[132] He explains that Grapte, who instructs the widows and orphans, is the "bare letter of Scripture." This corresponds to the *flesh*, or primary sense, of Scripture indicated above, which "admonishes those readers whose souls are in the stage of childhood." Origen also connects Grapte's instruction to the second sense of Scripture, the *soul*, those who are "no longer consorting with the unlawful bridegroom, but remaining in a widowed state because not yet worthy of the true Bridegroom."[133] Clement, on the other hand, is the reader who "has got beyond the letter" and instructs those who "have escaped from the bodily desires and lower aims."[134]

Origen is clear in his insistence that the spiritual reading, that which is "beyond the letter," is the most profitable, but it is not for everyone. That is why he does not reject either the fleshly/primary or the soul senses. "That we may profit by the primary sense of Scripture, even if we go no further, is evident from the multitudes of true and simple-minded believers."[135] Still, Origen urges this kind of reader to push on to the "soul" interpretation. As an example of this level of reading, he points to what Paul states in 1 Corinthians 9.[136] This kind of reading,

---

[131]Ibid. Italics in Lewis's translation.

[132]*Shepherd of Hermas*, Vision 2.4: "Therefore you will write two little books, and you will send one to Clement and one to Grapte. Then Clement will send it to the cities abroad, because that is his job. But Grapte will instruct the widows and orphans. But you yourself will read it to this city, along with the elders who preside over the church." Greek text and English translation can be found in *The Apostolic Fathers: Greek Texts and English Translations*, 3rd ed., trans. Michael W. Holmes (Grand Rapids: Baker Academic, 2007).

[133]*Philocalia* 1.11 (Lewis, 11-12).

[134]Ibid.

[135]Ibid., 1.13 (Lewis, 13).

[136]Ibid. (Lewis, 14). 1 Cor 9:9-10: "'It is written,' [Paul] says, 'Thou shalt not muzzle the ox when he treadeth out the corn. Is it for the oxen that God careth, or saith he it altogether for our

according to Origen, is offered by Paul "to suit the great body of believers, and [is] edifying for those who have no ear for better things."[137]
But it is the spiritual interpretation that is beyond these and most profitable. It is the spiritual interpretation that we must "everywhere seek
in a mystery." But for Origen, there is profit in all.

It is this profit to which Gregory of Nazianzus points in his funeral
oration for Basil. Gregory claims that in his threefold approach Basil
moves from "mere literal or symbolical interpretation to a still wider
view, as I proceed from one depth to another, calling upon deep after
deep, and finding light after light, until I attain the highest pinnacle."[138]
The Greek original communicates an important aspect that is missed
in the English translation. It indicates that "Basil leads us away from
seeing either *exclusively* literally or spiritually; he takes us further, from
depth to depth."[139] Gregory of Nazianzus praises Basil's exegesis, in part
because it avoids the exclusivity of either the literal or allegorical, not
merely because it moves from the literal to the allegorical. Given what
we saw in Origen above, Basil's approach is not much different.

This threefold scheme in Origen is important because of the connection to Basil. There appears to be good reason to doubt that Basil
has Origen specifically in mind in his comments about allegory. The
connection to Origen's well-established and accepted approach to
reading Scripture is too ingrained in Basil for him to reject his debt to
Origen in this area. This, of course, means that Basil's comments about
allegory must have some other referent.

In this vein, John McGuckin has offered a helpful context within
which we might understand Basil's remarks. He points out that
church fathers such as Basil had a different understanding of biblical

---

sake? Yea, for our sake it was written: because he that ploweth ought to plow in hope, and
he that thresheth, to thresh in hope of partaking'" (translation in ibid.).

[137]Ibid., 1.13 (Lewis, 14).

[138]Gregory of Nazianzus, *Oration* 43.67 (NPNF[2] 7:417-18).

[139]Bouteneff, *Beginnings*, 127. Italics in original. "*Mē mechri tou grammatos histasthai mēde blepein ta
anō monon alla kai peraiterō diabainein kai eis bathos eti chōrein ek bathous.*"

commentary than simply following the narrative of a text. "Their understanding of biblical exegesis," he explains, "remained that of the more ancient period, a more discontinuous, confessional, and event-centered typology on the Christ-event."[140] The determinative pattern for the church's reading of Scripture is found in Jesus himself, and the Christocentric principle that governs patristic exegesis for Basil and other Fathers is found, par excellence, in the narrative of the journey to Emmaus.[141]

Luke 24 narrates the story of Christ, immediately after his resurrection, meeting two men on the road to Emmaus. The two men were already discussing the resurrection when Jesus joined them. The story is well known – Jesus, "beginning with Moses and all the prophets . . . interpreted to them the things about himself in all the scriptures" (Lk 24:27). Later in the same chapter, Jesus is described as opening "their minds to understand the scriptures" (Lk 24:45). This text serves as a very good entry point into patristic understanding of the Bible because it identifies the centrality of Christ in biblical interpretation. The "Scriptures" that are referred to here are the Old Testament Scriptures, and the two men were trying to make sense of the death and resurrection of Jesus in light of them. But Jesus shows them that the key lies in him.

This Christocentric focus, however, did not lead the Fathers to explain how every scriptural passage has a reference to Jesus. Rather, it is "an understanding that certain key passages which are Christological types explain the authoritative Jesus-event, which in turn explains the re-establishment of communion with God, the covenantal foundation of the Church as the elect community, the sole principle which explains the existence of the covenant literature that typologically anticipated the Church."[142]

---

[140]McGuckin, "Patterns of Biblical Exegesis," 38.
[141]McGuckin offers many other passages besides Lk 24. These can be found in "Patterns of Biblical Exegesis," 38-39. I will detail here only the Emmaus passage for the sake of space.
[142]Ibid., 39.

This was the traditional proclamation of the church and was brought into clear focus by the apostle Paul himself: "I handed on to you as of first importance what I in turn had received: that Christ died for our sins in accordance with the scriptures, and that he was buried, and that he was raised on the third day in accordance with the scriptures" (1 Cor 15:3-4). McGuckin wonderfully labels this the "architecture" of Scripture. "In accordance with the Scriptures" did not mean that Paul searched and interpreted the entire biblical story per se. Rather, he directed his interests toward the clarification of the "Jesus salvation-event."[143] Biblical history is thus the medium through which Paul accomplished this clarification. This led him to understand biblical history as "foreshadowing" that Jesus salvation event. *Foreshadowing* is instructive here because it refers to the shadow cast that announces someone's imminent arrival. For Paul and the church fathers, "shadows do not illuminate; they adumbrate."[144] This is the classical Christian Scripture principle, and these early Christian exegetes (Paul and the Fathers) lift out certain aspects of the story and keep others "in the shade."[145]

We must appreciate the difference between this classic biblical and patristic approach and the typical approach of many evangelicals who have the assumption that fundamental norms can literally be pulled from any and every part of the biblical text.[146] I recall a board meeting a number of years ago in my own church. We were discussing the dwindling attendance at our services. During the discussion, appeal was made to Acts 2 and Pentecost. The well-meaning board member chastised our lack of vision for church growth by appealing to the final verse of Acts 2: "And day by day the Lord added to their number those who were being saved." The assumption on the part of this elder's appeal was that everything in Scripture was not only descriptive but also prescriptive and meant to be normative in any and all cases. But

---

[143]Ibid.
[144]McGuckin, "Patterns of Biblical Exegesis," 39n10.
[145]Ibid., 39.
[146]Ibid., 39n11.

this assumption that all of Scripture is strictly speaking a normative text is actually not supported by the apostolic tradition where Paul and Basil reside. The heart of Scripture for them is the mystery of Christ.

If we take what McGuckin states about the "architecture" and understand that Basil is working with these same assumptions, we begin to see how to understand his remarks on allegory. He thus raises a number of examples of the type we have already examined, where God's providential plan is shown in animals and compared to his expectations for humanity. Those who nurture feuds emulate the camel.[147] People who swallow up the small fry in business are like sharks.[148] People who change color with ease are like the octopus.[149] This series of comparisons to highlight God's providential plan reaches a climax when Basil speaks of the sea urchin that clings to rocks in rough weather: "If God has not put the sea urchin outside of His watchful care, does he not have regard for your affairs?"[150]

For McGuckin, all this should "contextualize properly what has often been lifted wholesale out of Basil and misinterpreted."[151] He is, of course, referring here to the comments on *Hexaemeron* 9.1. The passage is not a wholesale attack on allegorical reading of Scripture. Rather, it is against "excessive allegorization of details such that the moral import of the text's overarching *skopos* [architecture] loses itself in a welter of secondary details."[152] This has been labeled by some a translational type of allegory that has an arbitrariness about it and may actually lead to heresy.[153] It has no connection with the true spiritual meaning of Scripture, which resides in its architecture.

The referents who practice this translational allegory here in 9.1 and earlier in 2.2 are likely the Manichaeans.[154] Against them, Basil is trying

---

[147]Basil, *Hexaemeron* 8.1 (FC 46:118).
[148]Ibid., 7.3 (FC 46:111).
[149]Ibid. (FC 46:110).
[150]Ibid., 7.5 (FC 46:114).
[151]McGuckin, "Patterns of Biblical Exegesis," 45.
[152]Ibid.
[153]Bouteneff, *Beginnings*, 129; and Lim, "Politics of Interpretation," 357.
[154]Bouteneff, *Beginnings*, 129.

to assert the intentionality of Scripture. They are "counterfeiters of truth, who do not teach their minds to follow the Scripture, but distort the meaning of Scripture according to their own will."[155] The architecture of Scripture is the context within which we should understand Basil's accusations against certain allegorizers who consider themselves "wiser than the revelations of the Spirit" and introduce their own "ideas in pretense of an explanation."[156]

Just as the Christ salvation-event is key for the architecture of Scripture, the architecture of Scripture is key to understanding Basil's comments about allegory. We have seen that he not only "know[s] the laws of allegory" but endorses a deeper reading of Scripture as spiritually necessary.[157] When he refers in the passage at issue to the "common meaning of Scriptures," he juxtaposes it to the way some interpret "according to their own opinion" and "according to their own notions, just like the dream interpreters, who interpret for their own ends the appearances seen in their dreams."[158] In this context, the exhortation by Basil to let Scripture "be understood as it has been written" is not a call to attend to a literalistic attachment to the text but rather a call to attend to the purpose of Scripture wherein God "has ordained that all things be written for the edification and guidance of our souls."[159]

Given the general makeup of his audience as tradespeople, Basil understands that the final purpose of Scripture is the most important thing. These basic explanations on God's providence within cosmology are geared toward his audience. This accounts for the "plainness" of his homilies. Given that these comments on allegory were delivered during his last homily, it is almost as if he were apologizing for this plainness throughout.[160] In this vein Basil points out various scientific theories about the shape of the earth and how each one overthrows the

---

[155]Basil, *Hexaemeron* 2.2 (FC 46:22).
[156]Ibid., 9.1 (FC 46:136).
[157]Ibid. (FC 46:135).
[158]Ibid.
[159]Ibid. (FC 46:136).
[160]McGuckin, "Patterns of Biblical Exegesis," 45-46.

previous.[161] He labels the purveyors of these theories as employing "foolish wisdom." But his reason for this accusation is not because Moses' account trumps those scientific explanations. In fact, he concedes that Moses does not discuss them in Genesis because they are "useless for us, things in no way pertaining to us."[162] Scripture simply does not speak about these things in a scientific manner—this is not the architecture of Scripture. Basil claims value for "our version of creation" because they are the "words of the Spirit" that give us not scientific theories but "things . . . written for the edification and guidance of our souls."[163] He is critical of those who go beyond what Moses himself has written and give it a dignity on that basis.[164] Scripture needs to be "understood as it has been written" because adding to it with translational allegory or scientific theories goes beyond its scope and intent.

We thus have a number of related reasons for doubting that Basil is offering a wholesale rejection of allegory in his *Hexaemeron*. Origen was likely not the intended target of Basil's criticism of allegory. In fact, Basil employed the same way of reading Scripture as Origen. In this sense, one could say that he, as most church fathers, owed a certain debt to Origen for clarifying the exegetical principles for Scripture reading. This amounted to, first and foremost, the threefold method of interpretation. Basil knew it and approved of it. In fact, it is probably why Basil's homilies on the Hexaemeron avoid the spiritual/allegorical approach one might see in his homilies on the Psalms. The Hexaemeral homilies are literal because Basil knows his audience.

We know Basil received some heat from his audience for sticking to this literal reading. Knowing full well that his audience was primarily composed of tradespeople, he was employing Origen's threefold method in order to reach those to whom the primary/fleshly sense was

---

[161]Basil, *Hexaemeron* 9.1 (FC 46:135-36).
[162]Ibid. (FC 46:136).
[163]Ibid.
[164]Lim, "Politics of Interpretation," 359.

most profitable. Rather than outright rejecting the allegorical method, he is offering a warning to these spiritual novices not to abandon the literal meaning in favor of a spiritual meaning.[165]

Basil was employing a pastoral strategy out of concern for his audience. His purpose in the homilies was not to lead his hearers to contemplate creation in the "mystic-philosophic sense"[166] but to allow "the marvel of creation to gain such complete acceptance from you that, wherever you may be found and whatever kind of plants you may chance upon, you may receive a clear reminder of the creator."[167]

Recently, however, Stephen Hildebrand has called this conclusion based on Basil's audience into question.[168] Hildebrand points out that there is no indication that Basil's homilies on the Psalms were directed at an audience of more advanced Christians. This is a significant point because in these homilies Basil employs allegorical readings and explicitly refers to Origen's threefold hermeneutic.[169] If Basil's pastoral strategy to avoid allegory was based on his audience, why is allegory so central in his homilies on the Psalms, which had the same kind of audience?[170] Further, Basil's homily on Psalm 114 is delivered explicitly in a liturgical setting where one would certainly expect to see a variety of Christians present.

---

[165]Lim, "Politics of Interpretation," 360. See also Basil's Homily on Psalm 44: "The sayings of God have not been written for all, but for those who have ears according to the inner man" (*Homily* 17.2 [Way, 277]). See also his Homily on Psalm 45: "Having meditated on the expressions of the psalm in turn, you will learn the hidden meaning of the words, and that it is not the privilege of any chance person to gaze at the divine mysteries, but of him alone who is able to be a harmonious instrument of the promise, so that his soul is moved by the action of the Holy Spirit in it" (Homily 18.2 [FC:297-98]).

[166]Lim, "Politics of Interpretation," 362.

[167]Basil, *Hexaemeron* 5.2 (FC 46:69).

[168]Stephen M. Hildebrand, *The Trinitarian Theology of Basil of Caesarea: A Synthesis of Greek Thought and Biblical Truth* (Washington, DC: Catholic University of America Press, 2007), 137-38.

[169]A very good summary of the issue is presented by Adam David Rasmussen, "How St. Basil and Origen Interpret Genesis 1 in the Light of Philosophical Cosmology" (PhD diss., Catholic University of America, 2013), 77-83. My summary of the issue follows that of Rasmussen.

[170]Hildebrand (*Trinitarian Theology*, 137n177) quotes Jean Bernardi (*La predication des Pères cappadociens: le prédicateur et son auditoire* [Paris: Presses Universitaires de France, 1986], 33-34), who states that most of the homilies on the Psalms, if not all, have "au grand public" as their audience. Bernardi sees certain disparaging remarks from Basil about his audience's sins (*Hom. In Ps.* 29.3, 32.2).

The problem is that in each of these homilies Basil employs figurative interpretations.[171] Does this mean that the conclusion above, based as it is on Basil's pastoral strategy for his simple audience, cannot be accepted? Must this force us back to the conclusion that even though at one time Basil supported and employed Origen's interpretation strategy, by the time he gets to delivering his homilies on the six days of creation he has abandoned it in favor of a solely literal reading? Not necessarily. Basil can be understood as remaining squarely within Origen's interpretational strategy. In his *Hexaemeron* we do see Basil employing cautious use of figurative readings in his public homilies, but this is not the same as rejection of them.

Basil saw how heretics had put allegory to use. For example, he claims that the Marcionites, Valentinians, and Manichaeans distort the phrase in Genesis 1:2 "And Darkness was over the deep."[172] He states that darkness is taken in this verse to indicate an evil that opposes God. "If 'God is light' [1 Jn 1:5] they say, assuredly in conformity with the meaning, the power warring against Him would be darkness, a darkness not having its being from another, but a self-begotten evil."[173] Basil opposes this interpretation, not because it is allegorical in nature but rather because it would mean that darkness subsisted on its own and had not been made by God. This is why Basil, in this instance, claims that the meaning of this phrase is "simple and easily understood by all" and invites his listeners to "accept the concept of darkness simply and without curiosity."[174]

But just because Basil opposes an allegorical interpretation in this instance does not indicate a wholesale rejection of it and a turn to an exclusively literalistic approach. Just as accepting "darkness" as subsisting on its own is an affront to the priority of God as sole, eternal Creator, so is literalistic anthropomorphizing an affront to God's

---

[171]Rasmussen, "How St. Basil and Origen Interpret," 78.
[172]Basil, *Hexaemeron* 2.3 (FC 46:25-26).
[173]Ibid., 2.4 (FC:46:26).
[174]Ibid. (FC 46:27); and ibid., 2.5 (FC 46:29).

eternal nature. Thus, when he discusses God's speaking in Genesis 1:3, 6, he states,

> Let us first inquire how God speaks. Is it in our manner? . . . Does He thus manifest His hidden thought by striking the air with the articulate movement of the voice? Surely, it is fantastic to say that God needs such a roundabout way for the manifestation of His thoughts. Or, is it not more in conformity with true religion to say that the divine will joined with the first impulse of His intelligence is the Word of God?[175]

If one were to take this literally, one would have to conceive of God as corporeal. The issue appears not to be either allegorical or literal interpretation. Rather, there are other factors at work in determining how one ought to interpret.

Rasmussen explains that Basil saw heretical, figurative interpretations as a threat to his common audience because they could not discern them. This is why he chose not to employ them in his preaching on the six days of creation.[176] Even though his audience bemoans Basil's literal interpretation of the Hexaemeron, he refuses to capitulate to their desire for more allegory. Simple Christians have no need for it, and it may, in fact, lead some astray who lack discernment. The approach he takes as a result is well within Origen's hermeneutical theory because of its sensitivity to the levels of Scripture.

RECOMMENDED READING

Lim, Richard. "The Politics of Interpretation in Basil of Caesarea's 'Hexaemeron,'" *Vigiliae Christianae* 44, no. 4 (1990): 351-70.

Sandwell, Isabella. "How to Teach Genesis 1.1-19: John Chrysostom and Basil of Caesarea on the Creation of the World." *Journal of Early Christian Studies* 19, no. 4 (2011): 539-64.

---

[175]Ibid., 3.2 (FC 46:38).
[176]Rasmussen, "How St. Basil and Origen Interpret," 79.

5

CREATION OUT
of NOTHING

L anguage, according to George S. Hendry, is designed mainly for
speaking about things that are. We really have no way of speaking
about things that are not.[1] The confusion this causes is illustrated in an
exchange in Lewis Carroll's *Through the Looking Glass* between Alice and
the White King:

> "I see nobody on the road," said Alice.
> "I only wish *I* had such eyes," the King remarked in a fretful tone. "To
> be able to see Nobody! And at that distance too! Why, it's as much as I
> can do to see real people, by this light!"[2]

How does one speak about that which is not, when language is designed primarily for speaking of things that are?

Today we think we know exactly what *nothing* means. But is it really
as simple as we think? Consider the following attempt at defining *nothing*:

> We all speak freely of nothing – and think nothing of it. There seems to
> be no mystery to it. We do not have to consult the dictionary to find out
> what it means; we know, or think we know. Nothing is the polar opposite

---

[1]George S. Hendry, "Nothing," *Theology Today* 39 (1982): 274.
[2]Lewis Carroll, *Through the Looking-Glass and What Alice Found There*, Oxford's World Classics (New
York: Oxford University Press), 198-99. Italics in original.

of everything. This is how it appears in mathematics. Take a number, subtract from it one unit after another; the result is nothing: 12345 – 10000 – 2000 – 300 – 40 – 5 = 0. A child can do it – there is nothing to it.[3]

Because this paragraph uses the word *nothing* in a number of different ways and, therefore, with a number of different meanings, it illustrates the problem well. If, as the definition indicates, nothing is what remains after everything is taken away, then how can we properly describe it with the verb *is*? It is, in fact, illogical to attach the verb *is* to what remains after everything is taken away, because *is* signifies a property that belongs to its opposite.[4]

The problem with understanding and describing nothing was also alive in the ancient world and caused the idea of God creating out of nothing (*creatio ex nihilo*) to receive a fair bit of discussion. At issue is the argument put forward in 1978 by German scholar Gerhard May.[5] While May was not the first scholar to raise the issue, his extended treatment of it has received a wide hearing.[6]

May's argument is that, contrary to widespread belief, the Christian doctrine of God creating out of nothing was not present in the Old Testament and did not become standard in Christian theology until the latter half of the second century.[7] The widespread view, according to May, is that the doctrine was present in Hellenistic-Jewish theology, with 2 Maccabees 7:28 being offered as classic proof of its presence.[8]

---

[3]Hendry, "Nothing," 274.

[4]Ibid.

[5]Gerhard May, *Schöpfung aus dem Nichts* (Berlin: Walter de Gruyter, 1978). I will be using the English translation of this work, Gerhard May, *Creatio Ex Nihilo: The Doctrine of "Creation Out of Nothing" in Early Christian Thought*, trans. A. S. Worrall (1994; repr., London: T&T Clark, 2004).

[6]Paul Copan ("Is *Creatio Ex Nihilo* a Post-Biblical Invention? An Examination of Gerhard May's Proposal," *Trinity Journal* n.s. 17 [1996]: 77-79) provides a brief but helpful historiography of scholars before May who made similar arguments. See Ian G. Barbour, *Issues in Science and Religion* (Englewood Cliffs, NJ: Prentice-Hall, 1966); Arthur Peacocke, *Creation and the World of Science: The Brampton Lectures, 1978* (Oxford: Clarendon Press, 1978); and Langdon Gilkey, *Maker of Heaven and Earth: The Christian Doctrine of Creation in Light of Modern Knowledge* (Garden City, NY: Doubleday, 1959).

[7]May summarizes his position in *Creatio Ex Nihilo*, xi-xiv.

[8]2 Macc 7:28 (RSV): "I beseech you, my child, to look at the heaven and the earth and see everything that is in them, and recognize that God did not make them out of things that existed. Thus also mankind comes into being."

Since the doctrine was seen as already existent in Hellenistic-Jewish theology, it could be presupposed in the New Testament, and tracing its origins becomes a moot issue.

May refers to earlier scholars who had already cast doubt on this widespread view.[9] These scholars certainly agree that Hellenistic Judaism did talk about God creating "out of nothing," but this did not exclude the acceptance of an eternal material world, because the description was not meant in an ontological sense. "Thus, the proposition of the creation of the world 'out of nothing' does not have from the beginning the meaning that we quite naturally associate with it."[10]

In Christian theology, the doctrine as an ontological statement appeared only when it was intended to give expression to the omnipotence, freedom, and uniqueness of God in opposition to creation from unoriginate matter. This means that there was an alternative understanding in early Christian theology but that the ontological understanding of *creatio ex nihilo* won out as Christianity sought to discredit Greek understandings of creation from unformed matter. For May, this was really a matter of Christians essentially usurping the phrase with their own unique understanding. May states, "The opposite of the philosophical cosmology must of course be seen dialectically: the doctrine of *creatio ex nihilo* indeed breaks through principles of philosophical metaphysics, but it can only be articulated within the latter's frame of reference and by using its terms."[11] In order to understand the issue, the starting point for *creatio ex nihilo* must be within the context of the controversy of early Christianity with philosophy.[12]

---

[9]See May, *Creatio Ex Nihilo*, xi-xii.

[10]Ibid.

[11]Ibid., xii.

[12]I would be remiss if I did not note that although May's thesis has garnered wide support, not all have agreed with him. For a sampling of support, see Paul Blowers, *Drama of the Divine Economy: Creator and Creation in Early Christian Theology and Piety*, Oxford Early Christian Studies (Oxford: Oxford University Press, 2012), 167-84; and Frances Young, "'Creatio Ex Nihilo': A Context for the Emergence of the Christian Doctrine of Creation," *Scottish Journal of Theology* 44 (1991): 139-51. I should note that Young has some minor divergences from May's thesis that do not change its main thrust. For a sampling of disagreement, see Paul Copan and William

PHILOSOPHY

In the late sixth or early fifth century BCE, Parmenides penned the words that became axiomatic for all Greek philosophy:[13]

> How and whence grown? I shall not let you say or conceive, "from Not-being," for it cannot be said or conceived that anything is not; and then what necessity in fact could have urged it to begin and spring up later or before from Nothing? Thus it must either be entirely or not be at all. Nor will the strength of conviction ever impel anything to come to be alongside it from Not-being.[14]

Later, Aristotle (384–322 BCE) approved of and summarized the position of Parmenides and others that "what is cannot come to be (because it is already), and from what is not nothing could have come to be (because something must be present as a substratum)."[15] Lucretius (99–55 BCE) states this in a much more blunt way: "Nothing can be created from nothing."[16] Thus, there was a significant "weight of Greco-Roman philosophical cosmology against" understanding creation from nothing.[17]

Near the middle of the first century BCE, the abovementioned Roman poet and philosopher Lucretius wrote his famous philosophical poem *De rerum natura*, which in English is translated as *On the Nature of the Universe*. While we have very little biographical information about Lucretius, he himself tells us that he was a great admirer of the philosopher Epicurus (341–270 BCE) for turning traditional religion on

---

Lane Craig, *Creation Out of Nothing: A Biblical, Philosophical, and Scientific Exploration* (Grand Rapids: Baker Academic, 2004); and J. C. O'Neill, "How Early Is the Doctrine of *Creatio Ex Nihilo*?," *Journal of Theological Studies* n.s. 53, no. 3 (2003): 449-65.

[13]Robert Grant calls this a law of nature that was widely accepted in antiquity. See Robert M. Grant, *Miracle and Natural Law in Graeco-Roman and Early Christian Thought* (Eugene, OR: Wipf & Stock, 1952), 32.

[14]Parmenides of Elea, *On Nature* fr. 8.7-11. All quotations from *On Nature* are from Richard D. McKirahan and A. H. Coxon, *The Fragments of Parmenides: A Critical Text with Introduction and Translation, the Ancient Testimonia and a Commentary*, rev. and expanded (Las Vegas, NV: Parmenides Publishing, 2009).

[15]Aristotle, *The Physics* 1.8 (191a27-31).

[16]Lucretius, *On the Nature of the Universe* 1.154-55 (Melville, 7). All quotations from *On the Nature of the Universe* are from Lucretius, *On the Nature of the Universe*, trans. Sir Ronald Melville (Oxford: Clarendon Press, 1997).

[17]Blowers, *Drama of the Divine Economy*, 169.

its ear, so to speak. Traditional religion asserted that gods governed the world, but Lucretius argued (with Epicurus) that this belief is "crushed by the burden of religion" and that religion has a "ghastly countenance."[18] Lucretius credits Epicurus as being the first who raised "mortal eyes against [religion], first to take his stand against it" and "to break apart the bolts of nature's gates and throw them open."[19]

Epicurus and Lucretius were challenging a body of long-held views that are represented first by Platonists, later modified by Aristotelians, and then again modified by the Stoics.[20] Epicurus and Lucretius can be referred to as "Atomists" and their model of understanding the origin of the natural order as the "Infinite Universe" framework. Platonism, Aristotelianism, and Stoicism can be referred to as "Aristotelian" and their model as the "Closed World" framework.[21] It is here, in the "Closed World" framework, where Christianity would ultimately find its place, albeit for reasons of its own.[22]

The descriptions "Infinite Universe" and "Closed World" give us a clue as to the focus of each framework because *universe* and *world* are intended to indicate a distinction between the two. *Universe* refers to all there is. This means that there may be many worlds. *World* or *cosmos*, however, refers to an organized system of natural parts that is usually centered, metaphorically or literally, on the inhabited earth.[23]

David Furley offers a very helpful discussion of some major differences between the two frameworks in antiquity.[24] He explains that one may use two sets of antithetical descriptions to point out two significant

---

[18]Lucretius, *On the Nature of the Universe* 1.63-64 (Melville, 5).

[19]Ibid., 1.66-67, 70-71 (Melville, 5).

[20]David Furley, "The Greek Cosmological Crisis in Classical Antiquity," in *Cosmic Problems: Essays on Greek and Roman Philosophies of Nature*, ed. David Furley (Cambridge: Cambridge University Press, 1989), 225.

[21]These descriptive labels of the frameworks can be attributed to Alexander Koyré, *From the Closed World to the Infinite Universe* (Baltimore, MD: Johns Hopkins University Press, 1957); and David Furley, *The Greek Cosmologists*, vol. 1, *The Formation of the Atomic Theory and Its Earliest Critics* (Cambridge: Cambridge University Press, 1987), 1-2.

[22]Blowers, *Drama of the Divine Economy*, 23.

[23]Furley, *Greek Cosmologists*, 1:2.

[24]Unless otherwise noted, the following is drawn from Furley, *Greek Cosmologists*, 1:3-8.

differences. These antitheses are *evolution* and *permanence*, and *mechanism* and *teleology*.

EVOLUTION AND PERMANENCE

The "Infinite Universe" framework pictures all worlds, including our own, as subject to growth and decay. Writers in this school of thought attempted to describe different stages in this process, including the first formation of earth, water, air, and the fiery bodies in the heavens (the so-called great world-masses). Further, explanations were attempted to explain the origin of complex life forms and the return of the world to its precosmic condition. While we can legitimately call these attempts theories of *evolution*, we must be wary of equating them with modern biological theory because of the vast differences between them.

In the "Closed World" framework the world does not evolve in the sense described above; it is in some sense *permanent*. We can distinguish between the three varieties of this school of thought. Plato was understood to have taught that the world was created by a god. He may have not meant this literally, but this was how many in antiquity read him. Aristotle taught that the world has always been—that it had no beginning and no end. The Stoics taught that the world has a beginning and an end but that at the end it will be consumed by fire, only to return again to its original state and repeat its history in an endless cycle. Rather than adopting an evolutionary understanding of the natural things that are realized in the world, "Closed World" adherents saw them as *permanent*.

For Plato, the god who made the world used eternal forms as models for natural things. Aristotle argued that the kinds of minerals, vegetables, and animals—and the cosmos as a whole—were eternally the same. Stoics posited that a "seed formula" (*spermatikos logos*) for each natural kind was retained through periodic conflagrations that start each new world on the same path as the previous one. This is very different from the "Infinite Universe" framework, which argued that each natural kind found in the world must have a beginning because each world has a beginning in time. They had no patience for any type of eternal pattern or model that existed separately from the world, which meant that each natural kind was seen as growing from less complex elements.[25]

MECHANISM AND TELEOLOGY

Everything that happens in the natural world was explained *mechanistically* and reduced to matter in motion according to the "Infinite Universe" model. Change – that is, life – was reduced to the collision of particles of matter through space. For the "Closed World" understanding, this was not enough because it did not include the end or goal (*telos* in Greek) to which things were moving. This *teleology* was explained in two different ways. Aristotle described the purpose as an unexplained feature of natural objects. For Plato, and then later the Stoics, the purpose of nature was explained by a purposive planning mind. Every feature of the world was understood as having been chosen for the best by Providence, or by a good god. The differences between the two frameworks can be seen in the following table.

Atomists (Infinite Universe)	Aristotelians (Closed World)
accident	design (or at least order)
matter-in-motion explanations	purpose or teleological explanations
infinite universe	finite cosmos = universe
transient cosmos	eternal or repeating cosmos
our cosmos one of many	unique cosmos
matter and void	no void inside cosmos
atoms	continuum
linear dynamics	centrifocal dynamics
flat earth	spherical earth
material soul	immaterial soul (except Stoics)
evolution	creation, or eternity

Source: David Furley, "The Greek Cosmological Crisis in Classical Antiquity," in *Cosmic Problems: Essays on Greek and Roman Philosophies of Nature*, ed. David Furley (Cambridge: Cambridge University Press, 1989), 225.

[25]Ibid., 1:4.

In what follows I will highlight one representative thinker of each framework so we may see how some of these differences worked out "on the ground." Among the body of the "Closed World" views, Plato (427 BCE–347 BCE) is especially important because of his influence on Christian theology. Among the "Infinite Universe" views, Lucretius is important because of his connection to Epicurus. For our purposes, Plato and Lucretius will represent these two frameworks of creation in antiquity. It is important to understand the distinction at the outset of this chapter because these two models had tremendous influence on early Christian interpretation of origins in Genesis. In some cases the influence was reactionary, while in others it was more conciliatory. Since the description of each model can only be a summary, the focus will be on issues relevant to our topic.

PLATO

The dialogue *Timaeus* provides us with Plato's understanding of the "generation of the world."[26] Plato makes a foundational distinction between "being" and "becoming." These are really distinctions between two kinds of reality. The realm of being is the intelligible realm and is "apprehended by intelligence and reason"; it is "always the same."[27] The realm of becoming is that which is perceived by the senses and "without reason." Because it is always in the process of *becoming* it never really *is*. The distinction here is important for Plato because everything that becomes must, of necessity, have been created by some cause. Plato identifies this cause as "the father and maker of all this universe."[28]

When Plato uses expressions like "the father and maker of this universe," "God," "Mind," "Demiurge," "Architect," and "Craftsman," we should understand what he means. Basically, for Plato, anything that participates in being, as opposed to becoming, can be described as

---

[26]Plato, *Timaeus* 27 (Jowett, 447). All English translations of *Timaeus* are from Benjamin Jowett, *Dialogues of Plato*, vol. 3, 3rd ed. (London: Oxford University Press, 1892).
[27]Ibid. (Jowett, 448).
[28]Ibid., 28 (Jowett, 449). Plato later refers to the cause as "Mind" (47e) and "God" (53b).

divine – that is, anything belonging to the intelligible and immortal realm, including "all those beings whose affinity for eternal truth makes them directing powers within the universal order."[29]

After his explanation that "becoming" must have been created by the "father and maker of all," Plato explains that the pattern which this "artificer" had in view when he made the world was the realm of being.[30] An intelligible and unchanging model for "creation" was used, and a copy of this model was impressed on a "receptacle" that would nurture the creation.[31] Plato describes the chaotic state of the receptacle before it was brought into order.[32] The work of the creator was not really to create but to bring order out of disorder. The world is a copy of the eternal forms and is good because the creator was good. Plato adds that "God desired that all things should be good and nothing bad, so far as this was attainable."[33] He mentions the qualification of the creator's desire that all things be good as far as was attainable or possible because of the state of things before creation in Plato's conception. Indeed, this creation was from preexisting matter: "Wherefore also finding the whole visible sphere not at rest, but moving in an irregular and disorderly fashion, out of disorder [the creator] brought order, considering that this was in every way better than the other."[34]

This means that Plato's creator is not really free, because he was limited to matter that possessed certain properties and dictated the way in which he could use it.[35] The substance with which the creator worked

---

[29]Richard A. Norris, *God and World in Early Christian Theology: A Study in Justin Martyr, Irenaeus, Tertullian & Origen* (London: Adam & Charles Black, 1966), 20.

[30]Plato, *Timaeus* 29 (Jowett, 449).

[31]Ibid., 48e-53c (Jowett, 541-45).

[32]Ibid., 52d-53c (Jowett, 544-45).

[33]Ibid., 30 (Jowett, 450).

[34]Ibid. (Jowett, 450). See also Aristotle (*Physics* 1.8), who (agreeing with Parmenides) claimed that nothing could truly "come into being" unless it preexisted. He claims that it would be absurd to say that something could be created from nothing. See Richard Sorabji, *Time, Creation and the Continuum: Theories in Antiquity and the Early Middle Ages* (Chicago: University of Chicago Press, 1983), 245-47.

[35]John Zizioulas, "'Created' and 'Uncreated': The Existential Significance of Chalcedonian Christology," in *Communion and Otherness: Further Studies in Personhood and the Church*, ed. Paul McPartlan

did not have a starting point and thus the creator did not give the world its existence in the full ontological sense. Not only was Plato's creator limited by preexistent matter, he was also limited by the space in which the matter existed. This space possessed movement and expansion and thus change.[36] Thus, for Plato, creation

> is not an act by which god sets the world in motion from the start – since movement already existed as a property of space which itself pre-existed creation. Creation is rather an act which sets this movement in the right direction and draws from it a world which is as good (beautiful) as it can be in such conditions.[37]

The philosophical schools understood Plato's account of the material world as his deepest thoughts and used it as the basis of their theories.[38]

LUCRETIUS

When Lucretius starts to detail his explanation of the emergence of life in book 5 of *On the Nature of the Universe*, he begins with "In the beginning the earth gave birth."[39] This betrays his materialistic and mechanistic approach – one that moves toward his goal of not invoking the working of the gods.[40] So unimpressed is he with theories about divine involvement with creation that he states, "In no way for us the power of gods Fashioned the world and brought it into being; So great the fault with which it stands endowed."[41] He is explicit near the start of the poem by informing readers of his "first great principle . . . that nothing ever by divine power comes from nothing."[42] This first great

---

(London: T&T Clark, 2006), 250-51. See Plato, *Timaeus* 48a, where "Necessity" determines that the Craftsman is not open to changing or eliminating the properties. Only certain processes are allowed because Necessity limits what the Craftsman may do.

[36]Plato, *Timaeus* 50b-d.

[37]Zizioulas, "'Created' and 'Uncreated,'" 251n1.

[38]Frank Egleston Robbins, "The Influence of Greek Philosophy on the Early Commentaries of Genesis," *The American Journal of Theology* 16, no. 2 (1912): 220.

[39]Lucretius, *On the Nature of the Universe* 5.783-84 (Melville, 159).

[40]Ibid., 1.146-58 (Melville, 7-8).

[41]Ibid. (Melville, 142).

[42]Ibid., 1.149-50 (Melville, 7).

principle, which is repeated with fair regularity throughout the poem, sets the stage for all he addresses.[43] The gods have nothing to do with the earth. Thus, Lucretius denies not only divine creation but also providence and purpose in creation.[44] The gods have done nothing for humankind that we should bestow praise on them.[45]

Instead of invoking the gods, Lucretius attributes everything that happened in the creation of the world to matter in motion.[46] Matter consists of small solid-body particles that are invisible to the naked eye and eternal.[47] These are the "seeds or primal atoms of things from which now all creation has been made."[48] Each particle is indestructible and unaffected by any change except motion. These atoms exist and have motion in a void space that has no center or boundary. Change and compound structures occur when these particles of matter collide as they travel constantly downward through the void. Lucretius argued that if the motion of atoms was downward they could not collide and, therefore, no complex structures could be introduced. He thus introduced the theory of the swerve:

> While atoms move by their own weight straight down through the empty void, at quite uncertain times and uncertain places they swerve slightly from their course. You might call it no more than a mere change of motion. If this did not occur, then all of them would fall like drops of rain down through the void. There would be no collisions, no impacts of atoms upon atom, so that nature would never have created anything.[49]

---

[43]For example, ibid., 1.152-53: "Nothing can be created from nothing"; 2.177-81: "In no way for us the power of gods Fashioned the world and brought it into being"; 5.195-99: "In no way for us the power of gods Fashioned the world and brought it into being."

[44]Edward Peters, "What Was God Doing Before He Created the Heavens and the Earth?," *Augustiniana* 34 (1984): 63.

[45]Lucretius, *On the Nature of the Universe* 5.165 (Melville, 141).

[46]Furley, *Greek Cosmologists*, 1:4.

[47]Lucretius, *On the Nature of the Universe* 1.268-70 (Melville, 11).

[48]Ibid., 1.497-502 (Melville, 17). The Greek word *atomos* (invisible) is a compound word that comes from *a* (not) and *temno* (I cut), meaning uncuttable or indivisible, something that cannot be divided further. See Henry G. Liddell and Robert Scott, "ἄτομος," in *A Greek-English Lexicon*, www.perseus.tufts.edu/hopper/text?doc=Perseus%3Atext%3A1999.04.0057%3Aentry%3Da%29%2Ftomos.

[49]Lucretius, *On the Nature of the Universe* 2.216-24 (Melville, 42).

The motion of atoms, along with the swerve, has always been and will continue to be. Nothing can change it.[50]

For our purposes, the similarities and differences between Plato and Lucretius center on divine agency. Lucretius is explicit in his denial of any divine agency whereas Plato explicitly posits it. Both agree that the world was made (Plato), or grew (Lucretius), from preexisting matter. For Plato, the creator made order out of chaos, while Lucretius's mechanistic view points to atomic collisions that form compound structures. It should not be a surprise that the notion of divine agency also guides each philosopher's view of purpose. For Lucretius, there is no divine purpose, while for Plato divine purpose is important.

As we see above, one commonality between Plato's "religious" viewpoint and Lucretius's "scientific" perspective is the assertion of origins from preexisting matter. So deeply held was this conviction that it would have been absurd to anyone outside the Jewish and Christian traditions to believe otherwise.[51] In the remainder of this chapter we will look at a few select Fathers and their understanding of creation out of nothing. We see in them clear moves away from both of the aforementioned frameworks of creation and toward the formation of a uniquely Christian understanding. Theophilus is chosen because it is in him that we see the first clear assertion of the necessity of *creatio ex nihilo* in Christian theology. After Theophilus, it becomes commonplace and essential in all Christian theology in order to distinguish the Christian God from all else. I have chosen Ephrem and Basil as good examples of the more established doctrine of *creatio ex nihilo*.

THEOPHILUS OF ANTIOCH

Theophilus of Antioch is of special importance because he is usually identified as the first post–New Testament Christian writer to unam-

---

[50]Ibid., 2.294-307 (Melville, 44-45).

[51]Paul Blowers, "Doctrine of Creation," in *The Oxford Handbook of Early Christian Studies*, ed. Susan Ashbrook Harvey and David G. Hunter (Oxford: University of Oxford Press, 2008), 911.

biguously argue for *creatio ex nihilo* in the ontological sense. One can see that the doctrine is clearly set in opposition to the previously mentioned ideas concerning the eternality of matter.

Book 1 of Theophilus of Antioch's *Ad Autolycum* is his reply to Autolycus's request, "Show me your God."[52] Autolycus had apparently boasted about his gods who, according to Theophilus, cannot see or hear and are idols made by men. Theophilus proceeds to describe the Christian God, who is quite different. Throughout book 1 he contrasts the Christian God with the gods found in Hesiod, Homer, Plato, Euripides, and Orpheus.[53] It is likely that Theophilus studied these writers during his education, so he knew them well.[54] While Hesiod claims that muses inspired the writing of his *Theogony*, Theophilus claims the holy prophets, under the inspiration of the Spirit of God, as his authority for his doctrine of God.[55]

Book 2 begins by referring to the previous discussion in book 1. Theophilus finds it silly that artisans make gods (idols). While they are being made, they are considered by their makers as nothing at all. But then, when they are bought and sold and set up in a temple or home, they receive sacrifices and are worshiped. To Theophilus, this is "ridiculous" because he sees these gods merely as materials from which they were constructed. This is no different than what Autolycus is doing when he reads about the histories and genealogies of the "so-called gods." They were generated only when people read about them.[56]

These gods are not actually real. If they are, Theophilus reasons, surely they would continue to be generated up to the present day. Further, there should actually be more gods than men since gods do not die. He continues this line of thought, asking why Mount Olympus is now deserted,

---

[52]Theophilus of Antioch, *Ad Autolycum* 1.2 (Grant, 3). Cf. 1.14 (Grant, 21). All quotations from *Ad Autolycum* are from Theophilus of Antioch, *Ad Autolycum*, trans. Robert M. Grant, Oxford Early Christian Texts (Oxford: Clarendon Press, 1970).

[53]Carl Curry, "The Theogony of Theophilus," *Vigiliae Christianae* 42 (1988): 318.

[54]Grant, introduction to *Ad Autolycum*, xi.

[55]Hesiod, *Theogony* 36-38. Cf. Theophilus, *Ad Autolycum* 1.14 (Grant, 19).

[56]Theophilus, *Ad Autolycum* 2.2 (Grant, 23).

or why Zeus is no longer on Ida. Why is Zeus not everywhere instead of located in only one part of the earth? Contrast is made between these so-called gods and "the Most High and Almighty God," who is not confined in a place and "look[s] upon everything and hear[s] everything."[57] If this God or, by implication, another god could be confined to place, then the place containing him would be greater than he is. The Christian God is not thus contained but is "himself the locus of the universe."

Theophilus mentions both Epicurus (and by implication Lucretius) and Plato by name.[58] Epicurus denies the existence of God or, at best, thinks that God is interested in no one but himself. Plato and his followers do acknowledge that God is uncreated and the maker of the universe, but they also say that uncreated matter existed with God. To Theophilus this means that God is not the maker of the universe and his unique sovereignty is destroyed. Further, Theophilus believes that anything created is mutable and changeable, while anything uncreated is immutable and unchangeable. If matter is uncreated it must also be immutable and equal to God, who is uncreated and immutable. It is really unremarkable for God to create out of preexistent matter because even human beings can take material and fashion something out of it. "But the power of God is revealed by his making whatever he wishes out of the non-existent [*ex ouk ontōn*], just as the ability to give life and motion belongs to no one but God alone."[59]

When a person makes something, the limitations of the material affect the result. But God created whatever he wanted in whatever way he wanted. A human craftsman cannot give reason or breath or sensation to what she makes. God has an infinitely greater capability. He is not limited by the already existing matter, because he created it. He has the ability to create a being that is rational, breathing, and capable of sensation.

---

[57]Ibid., 2.3 (Grant, 25).
[58]Ibid., 2.4 (Grant, 27).
[59]Ibid.

Not only are the philosophers wrong in their conceptions of God, they are also in disagreement with other writers, particularly the poets.[60] Homer, for example, claims that the ocean is the origin of the gods.[61] But the ocean is just water and not God. In fact, God is the Creator of the universe and therefore of the ocean as well. Hesiod claimed that the world was created but did not say by whom. He also mentions a number of gods, but how can they be gods or know anything about the creation of the world if they themselves originated later than it?

Throughout *ad Autolycum* Theophilus takes particular aim at Hesiod's account of creation.[62] In 2.6 he explains that Hesiod assumes the pre-existence of matter at the origin of the world. The earth, according to Hesiod, was shaped by Chaos, which rules over gods and men.[63] But Theophilus asks, if Chaos reshaped the already existing matter and transformed it, who shaped and arranged it in the first place? Surely it could not have done this itself. Nor could the gods already mentioned since they were not in existence yet.[64] "Was there not instead a sovereign principle that made matter – I mean God, the one who set it in order?" Hesiod posits gods originating from earth and heaven and sea, and certain terrible men who were descendants of these gods. This, according to Theophilus, is "absolute nonsense."

---

[60]Ibid., 2.5 (Grant, 27-29).

[61]*Iliad* xiv.201, xxi.196.

[62]Peter C. Bouteneff, *Beginnings: Ancient Christian Readings of the Biblical Creation Narratives* (Grand Rapids: Baker Academic, 2008), 69; Frances Young, "Greek Apologists of the Second Century," in *Apologetics in the Roman Empire: Pagans, Jews, and Christians*, ed. Mark Edwards, Martin Goodman, and Simon Price (Oxford: Oxford University Press, 1999), 97-98; and Frances Young, *Biblical Exegesis and the Formation of Christian Culture* (Peabody, MA: Hendrickson, 2002), 55-57.

[63]Theophilus, *Ad Autolycum* 2.6 (Grant, 29-32), here cites Hesiod, *Theogony* 116-23, 126-33: "First Chaos came to be, and then / Wide-bosomed earth, ever-sure foundation of all / The immortals, who hold the peaks of snowy Olympus, / And dim Tartarus, the depth of broad-pathed Earth, / And Eros, fairest among the immortal gods / Who unnerves the limbs and overcomes the mind / And wise counsel of all gods and all men. / From Chaos came Erebus and black Night. / And Earth first bore, equal to herself, to cover her everywhere, / Starry heaven, that there might be an ever-sure abode for the / blessed gods; / And she brought forth long hills, graceful haunts / Of goddess Nymphs who dwell in woody hills; / She bore also the fruitless deep with his raging swell, / Pontus, without sweet union of love; and afterwards / She lay with Heaven and bore deep-swirling Ocean."

[64]Theophilus, *Ad Autolycum* 2.6 (Grant, 31).

Theophilus continues to mention names and stories about Pluto, Poseidon, and the sons of Zeus. Aristophanes's work called *The Birds* is also mentioned and quoted at length. All of these genealogies show the obvious – they are genealogies of men, not of gods. In a quasi-summary of all the names and stories mentioned thus far, Theophilus claims that there is no real need to keep citing these historians, poets, and philosophers because they are all deceived, as are those who pay attention to them.[65] Their myths do not really show them to be gods, only men – and some of them not very good men at that![66] In addition, the historians, poets, and philosophers made evil and inconsistent statements about the origin of the world. Some claimed that the world was uncreated and that nature was eternal. Others disagreed that it came into existence at all. The problem with all of these, however, is that their claims were made "by conjecture and human thought, not in accordance with the truth."[67] Some believed in providence whereas others did not. Whom is one to believe in these contradictions? The contradictions show that they do not know the truth and may even be inspired by demons.

In contrast to those who spoke through a spirit of error, the prophets of God were inspired and instructed by God himself. They were God's instruments, and through Wisdom they spoke about the creation of the world.[68] The first thing the prophets taught was that God made everything out of the nonexistent (*ex ouk ontōn*).[69] God existed "before the ages" (Ps 54:20 LXX), nothing was coeval with God, and he lacked nothing. God called this world out of nonexistence and willed to make humanity so that it might be known by him.

Theophilus connects the Wisdom through which the prophets spoke to the Logos through whom heaven was created. The Logos was "innate in [God's] own bowels [cf. Ps 109:3], [God] generated him together with

---

[65]Ibid., 2.8 (Grant, 35).
[66]Ibid. "Some of them drunkards, others fornicators, and murderers."
[67]Ibid.
[68]Ibid., 2.9 (Grant, 39).
[69]Ibid., 2.10 (Grant, 39).

his own Sophia [Wisdom], vomiting him forth [Ps 44:2] before every-thing else." This Logos was God's "servant" in creation, and through him all things were made. The Logos is also called "Beginning" because "he leads and dominates everything fashioned through him." But "Be-ginning" (Gen 1:1) and "Sophia" (Prov 8:22) are not the only appellations Theophilus attaches to the Logos. He is also called "Spirit of God" (Gen 1:2) and "Power of the Most High" (Lk 1:35). This is who came down to the prophets and spoke through them about creation. These prophets did not exist when the world was created, but God's Sophia and Logos were present. In support, Theophilus cites "Solomon" in Proverbs 8:27-29.[70] Further, it was the Logos who spoke through Moses, many years before Solomon, in Genesis 1:1: "In the Beginning God made heaven and earth."

Wisdom knew beforehand that many would speak nonsense about many nonexistent gods. Genesis 1:1 was recorded so that the real God may be made known through his works and to show that it was through his Logos that God made heaven and earth and all that is in them. Theophilus stresses that these are the first teachings given in divine Scripture. The emphasis, therefore, is "that the matter from which God made and fashioned the world was in a way created, having been made by God."[71]

Theophilus returns to Hesiod a few pages later, near the beginning of his discussion of the days of Genesis. He characterizes Hesiod's de-scription of creation through already existing matter as "quite feeble in relation to God."[72] Starting from materials already on the earth is really no different than what a human being can do. Man begins where he can, from below, from what is on the ground. The foundation must be laid first before the roof can be put on. This is not so with God, who "makes existent things out of the non-existent [*men ex ouk ontōn poiē ta*

---

[70]Ibid., 2.10 (Grant, 41), as cited by Theophilus: "When he prepared the heaven I was with him, and when he made strong the foundations of the earth I was with him, binding them fast."

[71]Ibid.

[72]Ibid., 2.13 (Grant, 47).

ginomena]," whatever he wills.[73] This is precisely why the prophet began by speaking of the creation of heaven, which was fashioned like a roof (Gen 1:1). God's power is shown by his creating the "roof" before he creates the foundation. Thus, "Things impossible with men are possible with God."[74] Ultimately, Theophilus desires to indicate the role of the Logos in creation. When Genesis 1:1 states "In the beginning God made heaven," it really means to say "*through* the Beginning heaven was created, as we have just explained."[75]

Theophilus puts forth this argument for creation out of nothing for three reasons: First, if matter is preexistent (and therefore uncreated), this places matter on the same level as God, and he can no longer be thought of as the Creator of everything; second, since God is uncreated he is, therefore, immutable (similarly, if matter was uncreated it would also be immutable); third, God creating out of preexistent matter is no different than a human craftsman fashioning something out of a given material.[76] Because God creates out of nothing and fashions "whatever he wished in whatever way he wished," he can, unlike human craftsmen, create a being that is rational, breathing, and capable of sensation.[77] In other words, for Theophilus creation out of nothing must be seen in parallel with the conferring of life.

EPHREM THE SYRIAN

In his *Commentary on Genesis* Ephrem is concerned that none of the Genesis narrative is misinterpreted to read of anything self-subsisting except God. This concern is manifested by his opposition to the teachings of Bardaisan. In fact, in their general introduction to Ephrem's

---

[73]Ibid. Earlier in 1.4 (Grant, 7), Theophilus makes a similar claim by citing 2 Macc 7:28: "*God made everything out of what did not exist [kai ta panta ho theos epoiēsen ex ouk ontōn eis to einai]*, bringing it into existence so that his greatness might be known and apprehended through his works."

[74]Theophilus, *Ad Autolycum* 2.13 (Grant, 47); cf. Lk 18:27.

[75]Ibid., italics mine; cf. ibid., 2.9-10 (Grant, 39-40).

[76]May, *Creatio Ex Nihilo*, 160.

[77]Theophilus, *Ad Autolycum* 2.4 (Grant, 27).

Commentary on Genesis, Mathews and Amar appeal to earlier independent studies that confirm "Ephrem's entire account of the six days of creation, on the surface a very literal commentary, is a polemic aimed primarily against the teachings of Bardaisan."[78] Thus, Ephrem's argument in his *Commentary on Genesis* must be set against the backdrop of his opposition to Bardaisan, whose ideas are very similar to many explained earlier.

Bardaisan (154–223 CE) became a Christian during the last quarter of the second century but had later separated himself from the church at Edessa.[79] He is regarded as a heretic by later ages. Neither his own writings nor even most of the confutations of his doctrines were preserved by monastic libraries. This means that his teaching requires reconstruction from scattered notices and accounts of later writers.

Ephrem's concern with Bardaisan's teaching centers on his cosmology.[80] Matter, Bardaisan argued, was divided into four pure "elements" or "self-subsisting beings," each of which existed in its own region: wind in the West, light in the East, water in the North, and fire in the South. These four self-subsistent beings were understood to be in conflict with a fifth, impure one – darkness, which occupied the lower regions. God, the Creator who occupied the region above, was seen as the sixth independent being and was in conflict with all five. Through some sort of mingling of these eternal elements chaos resulted. In this scheme, creation is an ordering of this chaos by God. For Ephrem, because these beings were self-subsistent, this makes God an

---

[78]Edward G. Mathews and Joseph P. Amar, "General Introduction," in St. Ephrem the Syrian, *Selected Prose Works*, trans. Edward G. Mathews Jr. and Joseph P. Amar, FC 91 (Washington, DC: Catholic University of America Press, 1994), 60.

[79]For more biographical information on Bardaisan, see L. W. Barnard, "The Origins and Emergence of the Church in Edessa During the First Two Centuries A.D.," *Vigiliae Christianae* 22 (1968): 169-73. This paragraph is based on Barnard's article.

[80]The following summary of Bardaisan's cosmology draws on Tryggve Kronholm, *Motifs from Genesis 1–11 in the Genuine Hymns of Ephrem the Syrian with Particular Reference to the Influence of Jewish Exegetical Tradition*, Coniectanea Biblica Old Testament Series 11 (Lund: CWK Gleerup, 1978), 29-30; and Kathleen McVey, "General Introduction," in Ephrem the Syrian, *Hymns*, trans. Kathleen McVey, Classics of Western Spirituality (New York: Paulist Press, 1978), 61-62.

arranger and not the Creator. In this light, we can see why Ephrem's foundational position in his commentary establishes God as the only self-subsistent being and the Creator of everything else.[81]

Already in the commentary's prologue we see Ephrem address these Bardaisanite themes. Up until the generation of the Tower of Babel, Ephrem explains, the "Creator had been manifest to the mind of the first generations."[82] From the time of the tower to the time of Moses it was publicly taught that creatures were created. But when the Israelites went astray in Egypt they, along with the entire world, started losing the teaching that God created these things out of nothing. They started thinking that these substances were self-existent and began calling things that had been fashioned from this creation "gods." God willed to set this right through Moses since they had become confused in Moses' generation. This would, asserts Ephrem, prohibit "this evil tradition" from being communicated to the whole world.

So God sent Moses to the Egyptians to correct mistaken beliefs about the Creator and creation. The mighty works and miracles performed by God through Moses were done in order to give Moses the necessary credibility regarding what he was going to write down. Thus, after performing these mighty works and miracles in Egypt, Moses wrote about the substances that were created out of nothing in order to inform the descendants of Abraham that these substances are not self-existent beings. Further, the things that were made out of something and then erroneously worshiped as gods were shown by Moses to be false gods. But God is not among these many gods, he is One.

Moses also wrote about "the mysteries of the Son that were inscribed when creatures were created."[83] This was done by inscribing types of the Son in the figures and events of Genesis. The works of his staff,

---

[81]Mathews and Amar, "General Introduction," 60.
[82]Ephrem, prologue to *Commentary on Genesis* 2 (FC 91:67). All quotations of Ephrem's *Commentary on Genesis* are from St. Ephrem the Syrian, *Selected Prose Works*, FC 91.
[83]Ibid., 4 (FC 91:68).

Ephrem adds, signify "allegorical and symbolical meanings."[84] According to Ephrem, the "true commandments" had been forgotten, so Moses needed to write about them, "while adding those that were necessary for the infantile state of the [Jewish] people."[85]

It is only after explaining this that Ephrem indicates, "[Moses] then wrote about the work of the six days." Creation occurred through a "Mediator" who had equal skill as the Maker and was of the same nature. Then, indicating an awareness of the shift of Genesis 2:4, Ephrem states that Moses went back and wrote of things that he had left out and not written about in his first account. This kind of summary is continued up until the end of the prologue.

Ephrem begins his commentary proper with Genesis 1:1 and claims that the verse indicates creation of the substance of the heavens and the earth.[86] His intention is to emphasize that "they were truly heaven and truly earth." Nothing else was signified by the words *heaven* and *earth*. It is from this context that Ephrem's comment on the six days must be understood. "Let no one," he states, "think that there is anything allegorical in the works of the six days."[87] Thus what is claimed to be created was, for Ephrem, created.

We should be aware of a couple of things before we conclude that Ephrem's statement about allegory is a broad rejection of the practice. First, we know that Bardaisan was forced to allegorize the Genesis account of creation to make it fit with his cosmology.[88] This could be a statement directed at that kind of allegorization in particular and not allegory in general. In other words, Ephrem's claim against an allegorical reading of the six days of creation is not concerned with exegetical method per se but with the results of that method.[89] He has

---

[84]Ibid.

[85]Ibid. (FC 91:68-69).

[86]Ibid., 1.1 (FC 91:74), citing Gen 1:1: "In the beginning God created the heavens and the earth."

[87]Ibid. (FC 91:74).

[88]Mathews and Amar, "General Introduction," 62.

[89]Jeffrey Wickes, "Ephrem's Interpretation of Genesis," *St. Vladimir's Theological Quarterly* 52, no. 1 (2008): 48.

theological concerns with Bardaisan, and those concerns provide the impetus for his comments here. Second, we have already seen in the prologue that Ephrem makes reference to "the mysteries of the Son that were inscribed when creatures were created."[90] Types of the Son were inscribed in the figures and events of Genesis, and Moses' staff had allegorical and symbolic meanings.[91] While we would not want to draw a straight line of connection to Ephrem's understanding of allegory and the developed Alexandrian understanding, we certainly cannot conclude that he would advocate an exclusively literalistic understanding either.[92] In other writings Ephrem actually scolds those who interpret Scripture in a rigid literal way.[93]

Heaven and earth were created by God from nothing, and nothing existed before God created them. In a clear jab at the teachings of Bardaisan, Ephrem states:

> Therefore, it is evident that heaven and earth came to be from nothing because neither water nor wind had yet been created, nor had fire, light or darkness been given their nature, for they were younger than heaven and earth. These things were created things that came after heaven and earth and they were not self-subsistent beings for they did not exist before [heaven and earth].[94]

Thus, nothing else was created along with the heavens and the earth. We know this, Ephrem explains, because Moses does not tell us that anything else was created. If anything had been made, "Moses would have said so." Nor did Moses tell us the day in which spiritual

---

[90]Ephrem, prologue to *Commentary on Genesis,* 4 (FC 91:68).

[91]Ibid.

[92]McVey, "General Introduction," 48.

[93]Ephrem, *Hymns on Faith* 31.1; *Hymns on Paradise* 11.6-7; *Commentary on the Diatessaron* 22.3.

[94]Ephrem, *Commentary on Genesis* 1.2 (FC 91:75). See also 1.14 (FC 91:85): "Heaven, earth, fire, wind, and water were created from nothing as Scripture bears witness, whereas the light, which came to be on the first day along with the rest of the things that came to be afterwards, came to be from something. For when these other things came to be from nothing, [Moses] said, God created heaven and earth. Although it is not written that fire, water, and wind were created, neither is it said that they were made. Therefore, they came to be from nothing just as heaven and earth came to be from nothing." This quotation begins a section that thoroughly emphasizes these themes.

beings were created.[95] This is an interesting comment here because in *Hymns to the Nativity* 26.5, Ephrem indicates that angels had, in fact, been created by the second day. This shows us that Ephrem's overriding purpose in the commentary of "a polemic aimed primarily against the teachings of Bardaisan" means that he does not really need to clear up this detail.[96]

Two paragraphs after he makes the statement about spiritual beings, he comes to a similar conclusion about Moses not giving us the whole story, which requires some knowledge of the context. In 1.3.4, Ephrem indicates Moses' words in Genesis 1:2 were that the earth was "void and desolate." The void and desolation were older than the elements, which were created later. This is not to say that void and desolation were "something," but rather that "the earth, which does exist, was known [to exist] in something which does not exist, for the earth existed alone without any other being." This, too, is likely a reference to the teachings of Bardaisan, who taught that there existed eternal principles. Ephrem here is pointing out that "even something that may or may not have any existence in and of itself preceded those elements that Bardaisan calls . . . eternal principles."[97] This is also a comment on Bardaisan's understanding of space, which he argued was a material substance that contained and enclosed everything. For Ephrem, this was a restriction of God since he would also be enclosed in this substance.[98] It is apparent, then, that Ephrem is trying to establish the priority of God over all else – "The earth, which does exist, was known [to exist] in something which does not exist, for the earth existed alone without any other thing."[99]

---

[95]Ibid., 1.3 (FC 91:76).

[96]Mathews and Amar, "General Introduction," 60.

[97]See FC 91:76-77n32.

[98]See *S. Ephraim's Prose Refutations of Mani, Marcion, and Bardaisan*, vol. 1, *The Discourses Addressed to Hypatius*, trans. C. W. Mitchell (London: Williams and Norgate, 1912), xciv-xcvi; and *S. Ephraim's Prose Refutations of Mani, Marcion, and Bardaisan*, vol. 2, *The Discourse Called 'Of Domnus' and Six Other Writings*, trans. C. W. Mitchell, A. A. Bevan, and F. C. Burkitt (London: Williams and Norgate, 1921), iv-vii, cxxiii-cxxiv.

[99]Ephrem, *Commentary on Genesis* 1.3.2 (FC 91:76).

BASIL OF CAESAREA

Basil actually says relatively little about creation out of nothing. He begins by establishing God as the source of creation of visible things. But before going into details on the Genesis account, Basil wants his hearers to understand "who is speaking to us."[100] He establishes Moses as the author and gives a brief biography of him. Moses was adopted and reared by Pharaoh's daughter. Because of this he received a royal education from the wise men of Egypt. But rather than remain there, he "preferred to suffer affliction with the people of God rather than to have ephemeral enjoyment of sin."[101]

A key event in Moses' life, according to Basil, was the slaying of the Egyptian in Exodus 2. As a result of this, Moses was banished from Egypt and went to Ethiopia and devoted forty years to the contemplation of creation. There,

> He, who, already reached the age of eighty years, saw God as far as it is possible for man to see Him, or rather, as it has not been granted to anyone else according to the very testimony of God: "If there be among you a prophet of the Lord, I will appear to him in a vision, or I will speak to him in a dream. But it is not so with my servant Moses, who is the most faithful in all my house: for I speak to him mouth to mouth: and plainly, and not by riddles."[102]

This is the man who reports to us in Genesis the things he heard from God.

After having established the authority of the account in Genesis because Moses, the author, saw God face to face, Basil wonders aloud what he should say first about Genesis 1:1 – should he refute other accounts or should he proclaim "our truth"?[103] Refutation is useless, he concludes. Many accounts have been written by the Greeks, but each surpasses an

---

[100]Basil, *On the Hexaemeron* 1.1 (FC 46:3). All quotations from *Hexaemeron* are from Saint Basil, *Exegetic Homilies*, trans. Agnes Clare Way, FC 46 (Washington, DC: Catholic University of America Press, 1963).
[101]Ibid. (FC 46:4).
[102]Ibid.
[103]Ibid., 1.2 (FC 46:5).

older one whenever they appear. Thus, there is no need to refute them since they contribute to their own undoing.

Some accounts of material origins explain the beginning of the universe and the elements of the world, while others think that visible things consist of atoms, and invisible things of molecules and a very small space. In the end, however, these writers weave a spider's web of weak and unsubstantial beginnings of things. If they could just say, "In the beginning God created the heavens and earth and sea," they would be in much better shape. Their desire to produce an account of origins devoid of God has lead them to offer accounts based on chance.

But with the opening words of Genesis, Moses immediately posits God as Creator. As Basil states, "How beautiful an arrangement! He placed first 'the beginning,' that no one might believe that it was without a beginning. Then he added the word 'created,' that it might be shown that what was made required a very small part of the power of the creator."[104] It was by his will alone that the Creator brought the material world into existence. Thus, we should not imagine a world without beginning and without end.[105] This first statement in the "divinely inspired teaching" of Genesis is "the preliminary proclamation of the doctrine concerning the end and the changing of the world."[106]

Things started in time must also be brought to an end in time. It is foolish to think the world is coeternal with God. The pursuit of geometry, arithmetic, the study of solids, and astronomy are "very laborious vanity" if those who pursue them believe that the world is coeternal with God. This ascribes to a limited world the same glory that is attributed to God, "the limitless and invisible nature."[107] To those who teach this Basil applies Romans 1:21.[108]

---

[104]Ibid. (FC 46:6).

[105]Ibid., 1.3 (FC 46:7), citing 1 Cor 7:31 and Mt 24:35.

[106]Ibid.

[107]Ibid.

[108]Ibid., citing Rom 1:21-22: "They have become vain in their reasonings, and their senseless minds have been darkened, and, while professing to be wise, they have become fools." Implicit in those who teach these sorts of things are Aristotle, *On the Heavens* 2.1.283b; and Cicero, *On the Nature of the Gods* 1.14.

The doctrine of *creatio ex nihilo* in the early church was established in direct opposition to certain philosophical and scientific views that encroached on God's providence, sovereignty, and eternality. I have avoided any extended discussion on disagreements with Gerhard May's thesis simply because it deflects us from this book's main purpose. In the Fathers discussed in this book, creation out of nothing was well accepted and seen as a necessary counter to certain philosophical and scientific views of God.

RECOMMENDED READING

Copan, Paul, and William Lane Craig. *Creation Out of Nothing: A Biblical, Philosophical, and Scientific Exploration.* Grand Rapids: Baker Academic, 2004.

May, Gerhard. *Creatio Ex Nihilo: The Doctrine of "Creation Out of Nothing" in Early Christian Thought.* Translated by A. S. Worrall. London: T&T Clark, 2004.

6

THE DAYS *of* GENESIS

In the previous chapter we looked at how a few representative Fathers understood *creatio ex nihilo* against the backdrop of their context. Similarly, this chapter will seek to understand the "days" of Genesis 1. The main difference between this chapter and the previous one is that understanding the days has been a much more controversial topic than *creatio ex nihilo*.

THEOPHILUS OF ANTIOCH

After Theophilus discusses the first two verses of Genesis, he quotes the entirety of Genesis 1:3–2:3 as a way of introducing his discussion on the days.[1] Genesis 1:3 starts with the creation of light because light reveals the things being set in order in creation. Thus, "Light is the beginning of the creation."[2] But understanding God's work in six days (Hexaemeron) is exceedingly difficult. In fact,

> no man can adequately set forth the whole exegesis and plan of the days'
> creation, even if he were to have ten thousand mouths and ten thousand
> tongues. Not even if he were to live ten thousand years, continuing in

---

[1]Theophilus of Antioch, *Ad Autolycum* 2.11 (Grant, 41-45). All quotations from *Ad Autolycum* are from Theophilus of Antioch, *Ad Autolycum*, trans. Robert M. Grant, Oxford Early Christian Texts (Oxford: Clarendon Press, 1970). See also chapter five.
[2]Ibid. (Grant, 41).

this life, would he be competent to say anything adequately in regard to these matters, because of the *surpassing greatness* [Eph 1:19] and the *riches of the Wisdom of God* [Rom 11:33] to be found in this account of the six days just quoted.[3]

Many philosophers, historians, and poets have imitated the Hexaemeron and tried to write a narrative on the work of creation, only to fail because of the "absence of even the slightest measure of truth in their writings."[4]

Theophilus begins his comments about the days proper by referring, first of all, to the fourth day, when the luminaries came into existence.[5] In the Genesis account, plants and animals come into existence before these luminaries. The reason for this order is to counter the philosophers who say that things produced on the earth come from the stars and thus "set God aside."[6] The order of plants and seeds coming into existence before the stars silences those who argue that stars produced the things on earth because "what comes into existence later cannot cause what is prior to it."[7]

The luminaries of the fourth day contain "a pattern and type of a great mystery" for Theophilus.[8] The sun is a type of God and the moon is a type of man because the sun is much greater in power and brightness than the moon. The sun is always full and does not wane, just as God

---

[3]Ibid., 2.12 (Grant, 45).

[4]Ibid. (Grant, 47). Even though Theophilus disparages philosophers, historians, and poets in this section, he still alludes to Homer's account of the massed Achaen army, which Homer states he could not recount "had he even ten tongues, or ten mouths" (Homer, *Iliad* 2.489). See Andrew Louth, "The Six Days of Creation According to the Greek Fathers," in *Reading Genesis After Darwin*, ed. Stephen C. Barton and David Wilkinson (Oxford: Oxford University Press, 2009), 44.

[5]Gen 1:14-19: "And God said, 'Let there be lights in the dome of the sky to separate the day from the night; and let them be for signs and for seasons and for days and years, and let them be lights in the dome of the sky to give light upon the earth.' And it was so. God made the two great lights – the greater light to rule the day and the lesser light to rule the night – and the stars. God set them in the dome of the sky to give light upon the earth, to rule over the day and over the night, and to separate the light from the darkness. And God saw that it was good. And there was evening and there was morning, the fourth day."

[6]Theophilus of Antioch, *Ad Autolycum* 2.15 (Grant, 51).

[7]Ibid.

[8]Ibid.

"always remains perfect and is full of all power, intelligence, wisdom, immortality, and all good things."[9] Conversely, every month the moon wanes only to be reborn and "waxes as a pattern of the future resurrection."[10]

The fourth day is also important for Theophilus because the three days prior to the creation of the luminaries are "types of the triad of God and his Logos and his Sophia."[11] He adds man to this triad and connects it allegorically with the creation of the lights on the fourth day. Thus, the three days before the luminaries are types of the triad of God, Logos, Sophia, while the fourth is a type of man.[12] Man is in need of light, and with this light, he is added to the triad "so that there might be God, Logos, Sophia, Man."[13] This is the reason the luminaries came into existence on the fourth day.

But Theophilus is still not finished drawing out patterns and types from the creation of the luminaries on the fourth day. The disposition of the stars has correspondence to the arrangement and rank of those who keep the law and commandments of God. The brightest stars, which remain in place and are unswerving, exist in imitation of the prophets. The stars of secondary brightness are types of righteous people. But the planets, the stars "which pass over and flee from one position to another," are a type of the men who wander from God.[14]

Theophilus now moves on to the fifth day.[15] In a way similar to his approach to the fourth day, he sees patterns and types in the creation

---

[9]Ibid. (Grant, 53).

[10]Ibid. A pattern of the resurrection is also indicated in 1.13 and 2.14.

[11]Ibid.

[12]Robert M. Grant, "Theophilus of Antioch to Autolycus," *Harvard Theological Review* 40, no. 4 (1947): 236.

[13]Theophilus, *Ad Autolycum* 2.15 (Grant, 53).

[14]Ibid.

[15]Gen 1:20-23: "And God said, 'Let the waters bring forth swarms of living creatures, and let birds fly above the earth across the dome of the sky.' So God created the great sea monsters and every living creature that moves, of every kind, with which the waters swarm, and every winged bird of every kind. And God saw that it was good. God blessed them, saying, 'Be fruitful and multiply and fill the waters in the seas, and let birds multiply on the earth.' And there was evening and there was morning, the fifth day."

of the animals from the waters. Through these are "demonstrated the manifold wisdom of God."[16] The varied progeny and sheer numbers of these animals testifies to the wisdom of God. But at a deeper level, the water creatures "serve as a pattern of men's future reception of repentance and remission of sins" through baptism.[17]

Carnivorous birds and great fish are like greedy men and transgressors. Some keep the law of God and eat seeds from the earth whereas others eat the flesh of the weak and are transgressors. The righteous "do not bite or harm anyone but live in holiness and justice, but robbers and murderers and the godless are like great fish and wild animals and carnivorous birds. They virtually consume those weaker than themselves."[18]

On the sixth day God created the quadrupeds, wild animals, and land reptiles.[19] These animals are not given a blessing by God, because it was reserved for man, who would be created later. The quadrupeds and wild animals are also presented as a type of men "who are ignorant of God and sin against him and mind earthly things and do not repent."[20] This is contrasted with birds, who are a type of those who repent of their sins and live righteously. They "take flight in soul like birds, minding things above and taking pleasure in the will of God."[21] But not all birds do this. Flightless birds are like those who are ignorant of God, because they are unable to fly and "to run the upward course to the divine nature."[22]

---

[16]Theophilus, *Ad Autolycum* 2.16 (Grant, 53).
[17]Ibid.
[18]Ibid. (Grant, 55).
[19]Gen 1:24-26: "And God said, 'Let the earth bring forth living creatures of every kind: cattle and creeping things and wild animals of the earth of every kind.' And it was so. God made the wild animals of the earth of every kind, and the cattle of every kind, and everything that creeps upon the ground of every kind. And God saw that it was good. Then God said, 'Let us make humankind in our image, according to our likeness; and let them have dominion over the fish of the sea, and over the birds of the air, and over the cattle, and over all the wild animals of the earth, and over every creeping thing that creeps upon the earth.'"
[20]Theophilus, *Ad Autolycum* 2.17 (Grant, 55).
[21]Ibid.
[22]Ibid.

Quite apart from the types that have been occupying the majority of Theophilus's comments on the days, he now comments on the origin of evil. Because nothing created by God can originally be evil, he asserts that the sin of man made wild animals evil or poisonous; when man sinned these animals transgressed with him. He implies that humanity has dominion over these animals, and this, in turn, implies a responsibility in their transgression. "If the master of a house does well, his servants necessarily live properly; if the master sins, his slaves sin with him."[23] Thus, when man sinned, the animals were carried along in that sin. But when man returns to his created state and no longer does evil, the animals will also be returned to their original tameness.

When he gets to the creation of man on this sixth day, he notes the difficulty of understanding it and claims that divine Scripture gives a "summary mention of it."[24] When God said "Let us make man after our image and likeness," he was revealing the dignity of man. Everything else was made by a word, but the only creation worthy of his own hands was the making of man. The statement by God, "Let us make man after our image and likeness," was not made in the sense of God needing help. Rather, "Let us make" was said to his own Logos and Sophia. On the seventh day God "rested from the works that he had made."[25]

EPHREM THE SYRIAN

Although Ephrem does not devote a lot of attention to discussing the days of creation, there are some sections of interest. In his *Hymn on the Nativity* 26 he connects the days of creation with the incarnation of Christ.[26] Just as the year of Jesus' birth is the foundation of life because of its benefits toward humanity, the first day "in the beginning" is the

---

[23]Ibid.

[24]Ibid., 2.18 (Grant, 57).

[25]Ibid., 2.19 (Grant, 57).

[26]Ephrem, *Hymn on the Nativity* 26 (McVey, 205-9). All quotations from *Hymn on the Nativity* 26 are from Ephrem the Syrian, *Hymns*, trans. Kathleen E. McVey, Classics of Western Spirituality (New York: Paulist Press, 1989).

foundation of creation.[27] Events in the second year, like the visit of the
magi and the slaying of the innocents, are compared to the day light
was created, when darkness was torn away by the light.[28] Thus, "the
ray of our Redeemer's birth entered and tore away the darkness upon
the heart."[29]

At this point in the hymn Ephrem goes back to the first day and
begins a typological consideration of the significance of the days of
creation. The first day, which is the source and beginning, is a "type of
the root that germinated everything."[30] But greater than this day is
"our Redeemer's day" that germinated even more significant things:

> For his death is like a root inside the earth,
> His resurrection like the summit in heaven,
> His words [extend] in every direction like branches,
> And like His fruit [is] His body for those who eat it.[31]

The second day sings praise about the birth of the Son, who is identified by Ephrem as "the Second, the voice of the First."[32] In his *Hymns
of Faith* Ephrem refers to Jesus in various ways: "Son,"[33] "Offspring,"[34]
"First-born,"[35] "his Second,"[36] and "Voice."[37] He is explicit that Jesus
is the reference in Genesis 1:3:

> In the beginning, things-made
> Were created by means of the Firstborn.
> For "God said,
> 'Let there be light,'" and it was created.

---

[27]Ephrem, *Hymn on the Nativity* 26.1 (McVey, 206).
[28]Ibid., 26.2 (McVey, 206).
[29]Ibid., 26.3 (McVey, 206).
[30]Ibid., 26.4 (McVey, 207).
[31]Ibid.
[32]Ibid., 26.5 (McVey, 207).
[33]Ephrem the Syrian, *Hymns on Faith* 6.1.6; 8.7; 13.11; 14.9. All quotations from *Hymns on Faith* are
from St. Ephrem the Syrian, *The Hymns on Faith*, trans. Jeffrey T. Wickes, FC 130 (Washington,
DC: Catholic University of America Press, 2015).
[34]Ibid., 6.1.11; 2.7; 2.10; 11.6.
[35]Ibid., 6.6.2; 7.1; 12.1.11; 16.3.7.9.
[36]Ibid., 6.9.11.
[37]Ibid., 6.13.10.

To whom did he command,
When, look: there was nothing?
When he commanded the light,
He did not command it, "Be!"
But he said, "Let there be!"
Indeed, the word is different —
"Be" from "Let there be.[38]

He continues to explain in this hymn that God directed his creative commands to Christ:

Fully revealed is the truth
To the one who wishes to see it:
The six days that were created
Bear witness to the six sides:
They proclaim the four corners,
And the height and the depth.
He did not command works
To make themselves.
By means of one another, [works] were created:
The Father commanded with a voice,
The Son brought the work to completion.[39]

The Son is active in creation in carrying out the commands of God. Praise is given to the Son regarding creation because it was he who commanded the firmament into existence, divided the waters above, and gathered the seas below. Further, just as he separated the waters, he also separated himself from the angelic beings to come down to humankind.[40]

Next, Ephrem compares Christ, who "makes all things grow" and who "came down and became a Holy Blossom," to the third day on which flowers and blossoms grew. Christ grew up to adorn and crown the victorious.[41] In the same manner, the fourth day is used to praise

---

[38]Ephrem, *Hymns on Faith* 6.6 (FC 130:92-93).
[39]Ibid., 6.13 (FC 130:95-96).
[40]Ephrem, *Hymn on the Nativity* 26.5 (McVey, 207).
[41]Ibid., 26.6 (McVey, 207).

Christ, who created the pair of luminaries that blind and unseeing men worship. Jesus, the "Lord of the luminaries," actually came down and "like the sun He shone on us from the womb." In doing this he opened blind men's eyes and enlightened those who stray.[42]

The fifth day witnesses the creation of reptiles, including the serpent who "deceived and led astray our mother." This reference is to Eve, who is described as a "young girl" and who is subsequently typified by Mary, whose "innocent womb" bore the "Wise One" who "shone forth," exposed the deceiver, and crushed him.[43]

Day six contrasts the fates of Adam and the evil one, who "envied" Adam. Christ is called the "Medicine of Life" who "diffused Himself to both." But the fate of each was different when Christ was incarnated and "offered to them both." The evil one was destroyed, while the mortal lived.[44] Ephrem calls the seventh day the sabbath, which gives rest to living beings. The "Gracious One" is "untiring" in his care of humanity and animals. Although Christ took on the "yoke" of mortality and "was struck a servant's slap in court," "He broke the yoke upon the free."[45]

At this point Ephrem leaves the Genesis account of creation and speaks of days eight (Christ's circumcision and later Christian rejection of the practice) and ten (the perfection of the number because of its relation to the name Jesus) in relation to the infancy of Christ.[46] Of interest in this section is a statement made by Ephrem about Christ, "whose power turns Creation back again."[47]

In his *Commentary on Genesis* Ephrem continues to unpack Moses' account by making an apparent reference to time. There was a "length of that moment that followed" the void and desolation and, therefore, preceded the creation of the elements. The moment remains unspecified

[42]Ibid., 26.7 (McVey, 207-8).
[43]Ibid., 26.8 (McVey, 208).
[44]Ibid., 26.9 (McVey, 208).
[45]Ibid., 26.10 (McVey, 208).
[46]Ibid., 26.11-13 (McVey, 208-9).
[47]Ibid., 26.12 (McVey, 209).

here – Ephrem does not even make an explicit connection here between day one and day two. He simply moves on to commenting about those elements that were created after the void and desolation. In this context, he cites Genesis 1:2, "Darkness was upon the face of the abyss." Again, we must see this in the context of his issues with Bardaisan. For Ephrem, the order that Moses writes these things is meant to indicate priority. He includes both "darkness" and "the abyss" as the elements that were created after the void and desolation.

It is not surprising that Ephrem is not really interested in expounding on "the abyss of waters," because his issue with Bardaisan does not lie there.[48] Of the waters, he states: "But how was it created on the day on which it was created? Even though it was created on the day and at this moment, Moses does not tell us here how it was created. For now, we should accept the creation of the abyss as it is written, while we wait to learn from Moses how it was created."[49]

But the darkness receives more comment than the abyss about what it might be. Some, Ephrem explains, say that it was a cloud. But since the firmament was not yet created, this could not be. He recognizes the problems of positing a cloud as the darkness (Gen 1:2), but he still insists that this must be the explanation: "Because everything that was created was created in those six days, whether it was written down that it was created or not, the clouds must have been created on the first day, just as fire was created along with wind, although Moses did not write about the fire as he did about the wind."[50] Although Ephrem is clear about the days here, the focus has not been on temporality but rather on priority of creation in order to establish the falsity of Bardaisan's eternal principles and self-subsistent beings. They are not in keeping with what Moses tells us.

---

[48]Ephrem, *Commentary on Genesis* 1.4 (FC 91:77). All quotations of Ephrem's *Commentary on Genesis* are from St. Ephrem the Syrian, *Selected Prose Works*, trans. Edward G. Mathews Jr. and Joseph P. Amar, FC 91 (Washington, DC: Catholic University of America Press, 1994).
[49]Ibid.
[50]Ibid., 1.5 (FC 91:77-78).

In 1.5.2 Ephrem makes the following claim: "It was necessary that everything be known to have its beginning in those six days." The focus here, however, is not on what the days were or their temporality; rather, the priority is on maintaining the story. Thus, he asserts that since clouds are brought forth from the abyss of water, they must have been created together. He then refers to Elijah who "saw a cloud rising up out of the sea."[51] But it was not only this priority of substance that requires creation in those six days but also the need to render service. In other words, the clouds were created on that first night because they rendered service by not allowing light through from the upper heavens. "Clouds were spread over all of Creation on the first night and on the first day."[52] They were created and spread out in order to bring about "the requisite night."[53] According to Ephrem, the "shadow" of the clouds that were responsible for the darkness lasted "twelve hours." After that, light was created beneath the clouds and "dispersed their shadow that had been spread over the waters all night."

When Ephrem turns to the statement "the wind of God was hovering over the face of the waters" in Genesis 1:2, he actually rejects what was becoming a common understanding. "Some," he states, "posit that this is the Holy Spirit and, because of what is written here, associate it with the activity of creation."[54] This too can be attributed to his issues with Bardaisanite beliefs. In *Hymns Against Heresy* 55, Ephrem explains Bardaisan's teaching about the Holy Spirit bearing two daughters called "the shame of the dry land" and "the image of the waters." For Ephrem, this attribution to the Holy Spirit of creative and procreative powers is blasphemy. The nature of the Holy Spirit is "too pure and invisible" and simply cannot have a daughter bearing the name "image of the waters."

---

[51] 1 Kings 18:44. Ephrem also cites "Solomon" in Prov 3:20 ("By his knowledge the depths broke forth and the clouds sprinkled down dew").

[52] Ephrem, *Commentary on Genesis* 1.5.2 (FC 91:78).

[53] Ibid., 1.6 (FC 91:78).

[54] Ibid., 1.7 (FC 91:79).

In other words, the Holy Spirit cannot in any way be visible.[55] Thus, the connection of the Holy Spirit to the wind hovering over the waters in Genesis 1:2 makes Ephrem more than a little uncomfortable in a commentary that is directed against Bardaisan's teaching. Having excluded the interpretation that the wind is the Holy Spirit, Ephrem now continues discussing the wind as a natural phenomenon. He states that just as he inferred the creation of the clouds on the first day, he must also infer the breeze, "which is the service of the wind."

He now turns his attention to the light that came to be "at the dawn of the first day." He again states that the night lasted twelve hours and that the light was created between the clouds and the waters.[56] Since the clouds were keeping out the light of the heavens, this light was a creation, different from the light of the heavens.[57] Ephrem is quite clear here about the connection of the creation of light to days and, therefore, time: "The light remained a length of twelve hours so that each day might also obtain its [own] hours just as the darkness had obtained a measured length of time. Although the light and the clouds were created in the twinkling of an eye, the day and the night of the first day were each completed in twelve hours."[58] This is really the first extended indication that Ephrem views the days as twenty-four-hour days.

Ephrem continues his explanation of the light as a "bright mist over the face of the earth."[59] It was necessary that it spread over the face of everything because it had to chase away the darkness. Relying apparently on the Genesis account of the sun's creation on the fourth day, Ephrem claims that this light rendered service for three days.[60] It

---

[55]See Tryggve Kronholm, *Motifs from Genesis 1–11 in the Genuine Hymns of Ephrem the Syrian with Particular Reference to the Influence of Jewish Exegetical Tradition*, Coniectanea Biblica Old Testament Series 11 (Lund: CWK Gleerup, 1978), 43-44; and Edward G. Mathews and Joseph P. Amar, "General Introduction," in St. Ephrem the Syrian, *Selected Prose Works*, trans. Edward G. Mathews Jr. and Joseph P. Amar, FC 91 (Washington, DC: Catholic University of America Press, 1994), 79nn43-46.

[56]Cf. Ephrem, *Commentary on Genesis* 1.6 (FC 91:78); and 1.8 (FC 91:80).

[57]Cf. ibid., 1.5.2 (FC 91:78).

[58]Ibid., 1.8.2 (FC 91:80).

[59]Ibid., 1.8.3 (FC 91:81).

[60]Ibid., 1.9 (FC 91:81).

"served for the conception and the birth of everything that the earth brought forth on the third day."[61] Without explicitly mentioning the sun's creation on the fourth day, Ephrem asserts that it was in the firmament to ripen whatever was brought forth under that first light. In fact, it was from this first light and from fire, which were both created on the first day, that the sun and moon were created.

While we cannot doubt that Ephrem assumes the days of Genesis being twenty-four-hours long, it is important to note that this is, in fact, an assumption. In other words, there appear to be issues that are more important to him than proving the literalness of the days. For example, when commenting on the appearance of plants on the land, Ephrem asserts that grass, herbs, and fruit-bearing trees were all brought forth "at dawn." Further, although these plants were "only a moment old at their creation," the grasses appeared as if they were months old and the trees as if they were years old, with fruit budding on their branches.[62] But Ephrem also gives us a reason for this – that the grass would be food for the animals who would be created two days later and that the fruit would be food for "Adam and his descendants, who would be thrown out of Paradise four days later."[63] While there is a definite temporal chronology that Ephrem assumes, his emphasis is on the reason for the order.[64] That is, he shows a greater interest in prioritizing the story's coherence than with proving the actual literal history (in our understanding). This is characteristically Antiochene and substantially curbs our desire to make more out of the temporality than Ephrem himself does.[65]

Immediately following these comments on plants, Ephrem confirms his understanding of temporality when he cites Genesis 1:14, 16:

---

[61]Ibid., 1.9.2 (FC 91:81).
[62]Ibid., 1.22 (FC 91:89-90).
[63]Ibid., 1.22.2 (FC 91:90).
[64]Ephrem makes the same kind of argument with regard to the moon's actual age and its apparent age in 1.25.2 (FC 91:91-92).
[65]See chapter three.

And God said, "Let there be lights in the firmament of the heavens to separate the day from the night," that is, *"one to rule over the day and the other [to rule] over the night."* That [God] said, *"Let them be for signs,"* [refers to] measures of time, and *"let them be for seasons,"* clearly indicates summer and winter. *"Let them be for days,"* are measured by the rising and setting of the sun, and *"let them be for years,"* are comprised of the daily cycles of the sun and the monthly cycles of the moon.[66]

Ephrem takes this chronology so strictly that when he directs his attention to the creation of two great lights in Genesis 1:16, he claims that God did not create them in the evening, because that would change the night into day, and morning would be given priority over evening. So even here, in the midst of a clear reference to a strict understanding of the days, his point appears to be the issues of priority and service. This is also the point he makes in his concluding statements about the creation of cattle, reptiles, and beasts on the sixth day in Genesis 1:24: "as would be suitable for the service of the one who, on that very day, was to transgress the commandment of his Lord."[67]

Finally, in commenting on Genesis 2:1-2, Ephrem asks, "From what toil did God rest?" God, Ephrem asserts, does not grow weary and thus does not need rest.[68] God resting on the seventh day means that he "blessed and sanctified" it. "For it was given to them in order to depict by a temporal rest, which He gave to a temporal people, the mystery of the true rest which will be given to the eternal people in the eternal world."[69]

BASIL OF CAESAREA

Basil represents a significant step in theological sophistication compared to Theophilus, which is immediately evident by his connection of the days to the nature of time and how "beginning" should be

---

[66]Ephrem, *Commentary on Genesis* 1.23 (FC 91:90). Italics in original translation.
[67]Ibid., 1.27.2 (FC 91:93).
[68]Ibid., 1.32: "Thus heaven and earth were finished, and all their host. And God rested on the seventh day from all His work which He had done."
[69]Ibid., 1.33 (FC 91:96).

understood in Genesis 1:1. The world, he argues, did have its beginning "in time," but there was something that existed before this world and, therefore, before time.[70] "This was a certain condition older than the birth of the world and proper to the supramundane powers, one beyond time, everlasting, without beginning or end."[71] The reference here is to "rational and invisible creatures" that "fill the essence of the invisible world."[72] It is the world to which Paul refers in Colossians 1:16.[73] Genesis 1:1 is a reference to the material world, which was "added to what already existed" and is a training ground for the souls of humanity and a dwelling place that is suitable for all things that are subject to birth and destruction.[74]

This, of course, leads him into a short discussion on time. The nature of time is fleeting – the past has vanished, the future has not occurred, and the present is gone before we know it.[75] This is also the nature of all that was made. All created things lack stability because they grow or decay. When Moses writes about the generation of the world, he is trying to teach us that time is natural to things that change. "In the beginning he created" means "in this beginning according to time."[76] This beginning is not in reference to the world being the first thing to exist or be created. Moses is describing the beginning of the existence of "visible and sensible creatures after that of the invisible and spiritual."[77] The reason Moses does not tell us about the creation of the angelic realm is because it is outside of time.

---

[70]Basil of Caesarea, *Hexaemeron* 1.4 (FC 46:8). All quotations from *Hexaemeron* are from Saint Basil, *Exegetic Homilies*, trans. Agnes Clare Way, FC 46 (Washington, DC: Catholic University of America Press, 1963).

[71]Ibid., 1.5 (FC 46:9).

[72]Ibid.

[73]As cited by Basil: "'For in him were created all things,' whether visible or invisible, 'whether Thrones, or Dominations, or Principalities, or Powers,' or Forces, or hosts of Angels, or sovereign Archangels."

[74]Ibid.

[75]Ibid.

[76]Ibid., 1.5 (FC 46:10).

[77]Ibid.

But Basil is here faced with a theological problem – How does the eternal God, who is beyond time, create in time?[78] He does not want to speak about God creating in a way that would make him dependent on the world or limited in any way by it. God, after all, created from nothing, and he alone is "limitless and invisible nature."[79] Accordingly, he discusses the various ways in which *beginning* could be understood. A first movement could be called a beginning. Something could also be called a beginning that contributes to a greater result, like the foundation of a house, or as in the verse "The fear of the Lord is the beginning of wisdom."[80] Further, the beginning of a work by an artist is called art, and the good goal of an activity is often considered the beginning of actions.

The use of *beginning* in Genesis 1:1 could also take some of these variations of meaning into account. It is possible to speak of beginning in the sense of the first movement in time – that is, when the heavens and the earth were created. It is also possible to speak of beginning in the sense of "some systematic reason directing the orderly arrangement of visible things." This shows that there was purpose in creation – that is, as "a training ground for rational souls" and a "school for attaining knowledge of God because through visible and perceptible objects it provides guidance to the mind for the contemplation of the invisible."[81]

Basil thinks that the act of creation was "instantaneous and timeless" and that the "beginning is something immeasurable and indivisible."[82] In other words, there was no time lag between God's determination to create and the creation itself.[83] In fact, time did not exist before the

---

[78]Louth, "Six Days of Creation," 49.

[79]Basil, *Hexaemeron* 1.3 (FC 46:7).

[80]Ibid., 1.5 (FC 46:10), citing Prov 9:10.

[81]Ibid., 1.6 (FC 46:11). In support, Basil cites "the Apostle," who states in Rom 1:20: "Since the creation of the world his invisible attributes are clearly seen . . . being understood through the things that are made."

[82]Ibid., 1.6 (FC 46:11).

[83]M. A. Orphanos, *Creation and Salvation According to St. Basil of Caesarea* (Athens: G. K. Parisianos, 1975), 48. Cf. Basil, *Against Eunomius* 2.21.

creation of the world. This means that the actual beginning, which coincides with the world, is not time. This beginning initiates what follows, but is not part of it.[84] The beginning of time, Basil asserts, is not yet time, just as the beginning of a house is not yet a house or the beginning of a road is not yet a road. In fact, this sense of beginning is "not even the least part" of time.[85] Some do argue that the beginning is time. But this argument divides beginning into parts of time – the beginning, middle, and end. But surely it is ridiculous to speak of the beginning of a beginning. Further, dividing the beginning makes many beginnings. This beginning does not possess a beginning, middle, and end.[86] Moses recorded the words "In the beginning he created" in order to inform us that the world came into existence instantaneously at the will of God. Basil points to "other interpreters" of "In the beginning he created" who translated it as "'God made summarily, that is, immediately in a moment.'"[87]

The word *created* also carries significance with respect to *beginning* in Genesis 1:1. Basil refers to theoretical and practical skills. Theoretical skills involve the mind whereas practical skills involve the body. When the action involved in practical skills ends, nothing remains for those watching; the action ends with itself. But in the case of creative skills, even though the action ends, the work remains. For example, in architecture, carpentry, metal work, and weaving, even though the craftsman is absent, the works "manifest in themselves the artistic processes of thoughts, and make possible for you to admire the architect from his work."[88] Basil compares the world to a "work of art" through which we may contemplate the wisdom of him who created it. By the statement "In the beginning God created the heavens and the earth,"

---

[84]Louth, "Six Days of Creation," 49.

[85]Basil, *Hexaemeron* 1.6 (FC 46:11).

[86]John F. Callahan, "Greek Philosophy and the Cappadocian Cosmology," *Dumbarton Oaks Papers* 12 (1958): 35-36.

[87]Basil, *Hexaemeron* 1.6 (FC 46:11). The "other interpreters" here are Aquila, Symmachus, and Theodotion. See Orphanos, *Creation and Salvation*, 49.

[88]Ibid., 1.7 (FC 46:12).

Moses is indicating that the earth came into existence after the creation of the invisible world.[89]

Basil's concerns thus far are that creation is not coeternal with God and that he created the invisible world prior to the "beginning" of Genesis 1:1. This has been his focus in the first seven sections of his first homily on the Hexaemeron. In section eight he explains that an inquiry into creation, whether it is in the world of contemplation or the world that is open to our senses, would require a long, drawn-out account. He wonders if this is really necessary because "a concern for these things is not at all useful for the edification of the church."[90]

He calls his listeners to recognize that Isaiah gives us sufficient knowledge concerning the substance of the heavens: "'He established the heaven as if smoke,' that is, he gave the substance for the formation of the heavens a delicate nature and not a solid and dense one."[91] Further, with regard to the form of the heavens, Basil states that it is enough to read the words of Isaiah in praise of God: "He that stretcheth out the heavens as a vaulted ceiling."[92] Similarly, with regard to the earth, Basil claims that we should not delve into its substance or seek its foundation. Rather, we should be concerned with the reason it exists. "Therefore," Basil concludes, "I urge you to abandon these questions and not to inquire upon what foundation it stands. If you do that, the mind will become dizzy, with the reasoning going up to no definite end."[93]

In order to illustrate the endless reasoning required for this type of inquiry, Basil continues by summarizing some of the questions that may be asked of the nature and form of the heavens and the earth. Again, he concludes with a call to abandon these kinds of questions. This time the appeal is to Job 38:6: "Set a limit, then, to your thoughts, lest the words of Job should ever censure your curiosity as you scrutinize things

---

[89]Ibid.
[90]Ibid., 1.8 (FC 46:14).
[91]Ibid., citing Is 51:6 (LXX).
[92]Ibid., citing Is 40:22 (LXX).
[93]Ibid.

incomprehensible, and you also should be asked by him: 'Upon what
are its bases grounded?'" Even when Scripture appears to offer an explanation of these things, as in Psalm 74:4 (LXX), should we believe that
the sustaining force is pillars?[94] What about Psalm 23:2?[95] Basil asks,
"Now, how does water, which exists as a fluid and naturally tends to flow
downward, remain hanging without support and never flow away?"
How are we to understand this if what we know about water reasons
against this kind of claim?[96]

Apparently Basil does not consider these passages of Scripture as
scientific (literal) descriptions of the nature and form of the heavens,
and bringing this up in his homilies on the six days of creation may
serve as an important lesson for us. He therefore asks why, when people
hear non-scriptural arguments about the earth "suspended by its own
power," they do not question the legitimacy of that explanation by an
appeal to reason, "since it has a heavier nature."[97] He appears to accept
that we may say things like "the earth stands by its own power" or that
it "rides an anchor on the water." But we must recognize and not depart
from true religion, which ultimately claims that the Creator keeps all
things under control by his power. Basil appeals to Psalm 94:4 (LXX) to
indicate how the earth is supported: "'In the hand of God are all the
ends of the earth.' This is the safest for our own understanding and is
most profitable for our hearers."[98]

But this "safe" reply is not mere fideism. Basil proceeds to offer the
explanations "some of the inquirers into nature" give about the nature
and form of the earth. These are explanations drawn from Aristotle.[99]
But rather than attempting to refute the arguments, he accepts their
plausibility:

---

[94]Ibid., 1.9 (FC 46:15-16), citing Ps 74:4 (LXX): "I have established the pillars thereof."
[95]Ibid. (FC 46:16), citing Ps 23:2 (LXX): "He hath founded it upon the seas."
[96]Ibid.
[97]Ibid.
[98]Ibid.
[99]Aristotle, *On the Heavens* 2.13.295, 296a.

And should any of these things which have been said seem to you to be plausible, transfer your admiration to the wisdom of God which has ordered them so. In fact, our amazement at the greatest phenomena is not lessened because we have discovered the manner in which a certain one of the marvels occurred. But, if this is not so, still let the simplicity of faith be stronger than the deductions of reason.[100]

The same could be said about the heavens, says Basil. He appears to be much more dismissive of the arguments he lists of the nature and form of the heavens: "If we undertake now to talk about these theories, we shall fall into the same idle chatter as they."[101] But again, ultimately what matters to Basil, because Moses has stated it, is that "God created the heavens and the earth."

> Let us glorify the Master Craftsman for all that has been done wisely and skillfully; and from the beauty of the visible things let us form an idea of him who is more than beautiful; and from the greatness of these perceptible and circumscribed bodies let us conceive of him who is infinite and immense and who surpasses all understanding in the plentitude of his power. For, even if we are ignorant of things made, yet at least, that which in general comes under our observation is so wonderful that even the most acute mind is shown to be at a loss as regards the least of the things in the world, either in the ability to explain it worthily or to render due praise to the creator, to whom be all glory, honor, and power forever. Amen.[102]

Basil also has a short section on the days at the end of his second homily, which deals specifically with Genesis 1:2-5. Commenting on the statement "But the earth was invisible and unfinished," he claims that the earth was "incomplete" because plants, trees, flowers, and other things of the earth came only later "by the command of God."[103] So too, neither the moon, sun, nor stars had yet been made. There are two possible reasons why earth was called "invisible" by Scripture: because

---

[100]Basil, *Hexaemeron* 1.10 (FC 46:17).
[101]Ibid., 1.11 (FC 46:19).
[102]Ibid.
[103]Gen 1:2, as cited by Basil in ibid., 2.1 (FC 46:21). See also ibid., 2.1 (FC 46:22).

the first human had not been created to see it, or because everything on the earth was submerged under water and not illumined, causing it to be unseen. Of the two reasons, Basil opts for the second.

Basil's previously stated assumption about matter continues here – that God created all things out of nothing. He continues to take aim against those who distort the meaning of Scripture to say that matter is implied by the words of Genesis 1:2: "God, however, before any of the objects now seen existed, having cast about in His mind and resolved to bring into being things that did not exist, at one and the same time devised what sort of a world it should be and created the appropriate matter together with its form."[104] So, when Genesis states that "God created the heavens and the earth," it means that he made both the substance and the form.

Many things, however, were left unmentioned by Moses in Genesis 1:1 – in particular, water, air, fire, and the conditions produced from these that form an essential part of the world. The cosmological theory that these four elements (earth, fire, air, and water) through their union and mixture composed various objects of creation was generally accepted by the ancient philosophers and was prevalent during Basil's time.[105] But, according to Basil, even though these are not explicitly mentioned in the narrative, they were surely called into existence at the same time as the universe. No mention has been made in Genesis thus far about the creation of water. But since it was stated that the earth was "invisible," Basil claims we must deduce that "water abounded on the surface of the earth because the liquid substance had not yet been separated and spread in its allotted place."[106] This is also used to explain the unfinished nature of the earth since excess moisture can hinder the earth's productivity. The finished nature of the earth consists in its adornment with plants, trees, flowers, and other things, but the Genesis

---

[104]Ibid., 2.2 (FC 46:24).

[105]Orphanos, *Creation and Salvation*, 56. See Plato, *Leg.* X, 889B, 891C; and Aristotle, *De generatione et corruption*, 330b, 3-6.

[106]Basil, *Hexaemeron* 2.3 (FC 46:25).

account has not indicated the creation of any of these yet. But there is a "power stored up in [the earth] by the Creator," which waits for the divine command at the proper time to "bring its offspring into the open."[107]

Marcionite, Valentinian,[108] and Manichaean[109] distortions of Scripture also abound regarding the phrase in Genesis 1:2, "And darkness was on the face of the deep." Darkness is taken in these distortions as indicating evil that opposes God. "If 'God is light' [1 Jn 1:5] they say, assuredly in conformity with the meaning, the power warring against Him would be darkness, a darkness not having its being from another, but a self-begotten evil."[110] But this is unacceptable for Basil because it would mean that darkness had subsisted on its own and not been made by God. For Basil the words here in Genesis 1:2 are not as sinister as some would lead us to believe. In fact, their meaning is "simple and easily understood by all."[111] The "earth was invisible" because it had fathomless (i.e., deep) water covering it. No part of the earth showed through the waters because of the darkness above the water. This is the meaning of Scripture, and Basil calls his listeners to "accept the concept of darkness simply and without curiosity."[112] The call to accept Scripture "simply and without curiosity" must be understood within the context that Basil offers here — against the teachings of the Marcionites, Valentinians, and Manichaeans.

Since darkness is simply the absence of light, Basil asks about the absent light resulting in darkness on the face of the water. He admits that he must make an inference about the things that occurred before the formation of this "perceptible and destructible world."[113] The

---

[107]Ibid., 2.3 (FC 46:26).

[108]For Gnostic groups, a demiurge originated the world. This "creator" was a separate, fallen deity formed by emanations from the supreme Divine Being and, therefore, inferior. Since the universe is the work of the fallen demiurge, it is evil, imperfect, and always has an antagonistic relationship with the spiritual world. See Orphanos, *Creation and Salvation*, 40.

[109]The fundamental idea of the Manichaeans was the eternal contrast between good and evil, darkness and light. For them, the earth and the visible heavens were formed from the dismembered parts of the evil demons of darkness and were, therefore, evil by nature. See Orphanos, *Creation and Salvation*, 40.

[110]Basil, *Hexaemeron* 2.4 (FC 46:26).

[111]Ibid. (FC 46:27).

[112]Ibid., 2.5 (FC 46:29).

[113]Ibid.

ranks of angels and heavenly armies were certainly not in a condition of darkness but rather were "in light and in all spiritual gladness."[114] In the previous homily he already made reference to "rational and invisible creatures" that "fill the essence of the invisible world."[115] They belong to a world that is "beyond time, everlasting, without beginning and end" and which, of course, existed before the birth of this world – that is, before any description of creation in Genesis 1:1. He continues in this vein and cites several verses that show "heavenly light among the promised blessings."[116] Thus, "Solomon says: 'The just shall have light eternal'; and the Apostle says: 'Rendering thanks to God the Father, who has made us worthy to share the lot of the saints in light.'"[117] The darkness/light distinction is then applied to the "damned" who "are sent 'into the darkness outside,'"[118] and "those who have performed acts deserving of approbation have their rest in the supramundane light."[119]

The darkness amid this heavenly light is compared to a tent that is set up at midday.[120] The tent is made of material that completely blocks out light – this is the same kind of darkness that Genesis 1 indicates. It does not subsist on its own but results from shutting out the light of the heavens. When God commanded the physical heavens into existence and the heavens surrounded "completely the space enclosed by their own circumference with an unbroken body capable of separating the parts within from those outside, necessarily they made the regions within dark, since they had cut off the rays from the outside."[121] Since water covered the surface of all things, Genesis indicates that darkness was on the surface of the deep.

---

[114]Ibid.
[115]Ibid., 1.5 (FC 46:9).
[116]Ibid., 2.5 (FC 46:29).
[117]Prov 13:9 (LXX); Col 1:12.
[118]Mt 22:13.
[119]Basil, *Hexaemeron* 2.5 (FC 46:30).
[120]Ibid.
[121]Ibid.

For Basil, and contrary to Ephrem, the Spirit of God stirring over these waters is a reference to the Holy Spirit, "which forms an essential part of the divine and blessed Trinity."[122] *Stirring* here indicates a fostering care, or "preparing the nature of water for the generation of living beings." Thus we see that the Holy Spirit is also part of the creation process.

It was the command of God that did away with the darkness – "God said, 'Let there be light.'" This command, however, was the "divine word . . . not a sound sent out through phonetic organs, nor air struck by the tongue."[123] Rather, God's will is presented in the form of a command because this is easily understood by those who are being instructed. Similarly, the pronouncement in Genesis 1:4 that "God saw that the light was good" does not indicate a physical seeing by God but rather the "future advantage" of light.[124]

When Genesis 1:4 states "And God separated the light from the darkness," it means that they were divided by God and their natures made incapable of mixing. "God called the light Day and the darkness he called Night."[125] Basil cannot avoid speaking about the sun, even though it had yet to be created in the Genesis narrative. It is day when the earth is illuminated by the sun, and night when the sun is hidden, causing darkness. But Basil points out that at this point in the narrative it was not solar motion that caused the light and the darkness. Rather, it was "when that first created light was diffused and again drawn in according to the measure ordained by God, that day came and night succeeded."[126]

Later, in his sixth homily, Basil comments on this separation of light from darkness before the sun had been created. He cites Genesis 1:14-15 and gives a summary order of creation thus far:[127]

---

[122]Ibid. (FC 46:30).

[123]Ibid., 2.7 (FC 46:32).

[124]Ibid. (FC 46:32-33).

[125]Gen 1:5, cited in ibid., 2.8 (FC 46:33).

[126]Ibid.

[127]Gen 1:14: "And God said, 'Let there be lights in the dome of the sky to separate the day from the night.'"

The heavens and the earth had come first; after them, light had been created, day and night separated, and in turn, the firmament and dry land revealed. Water had been collected into a fixed and definite gathering. The earth had been filled with its proper fruits; for, it had brought forth countless kinds of herbs, and had been adorned with varied species of plants.[128]

Neither the sun nor the moon were in existence yet. Basil reasons that this is to avoid the possibility of identifying the sun as the "first cause and father of light" and to keep those ignorant of God from claiming it as "the producer of what grows from the earth."[129] Thus, there was a fourth day when God made two great lights "for the illumination of the earth."[130]

Why, Basil asks, is the sun said to have been made to give light if the creation of light had already occurred? He points out that the word used here is *illumination* (*phausis*) instead of *light* (*phōtismos*). "This does not conflict with what has been said about 'light' [*phōs*]." In Genesis 1:3 it was the nature of light that was created whereas here, in Genesis 1:14-15, the sun was created to be a "vehicle for that first-created light."[131] To explain this, Basil compares fire with a lamp. Fire has the power to give light whereas the lamp is to show light to those who need it. In the case of the Genesis account, "the lights have been prepared as a vehicle for that pure, clear, and immaterial light."[132]

In this discussion of light Basil again sees the need to draw parallels that one cannot classify as literal in the modern sense of the term. He refers to "the Apostle" who states "that there are certain lights in the world."[133] But there is also the "true Light of the world" in whom people participate. This participation in the true Light enables people to

---

[128]Basil, *Hexaemeron* 6.2 (FC 46:85).
[129]Ibid.
[130]Ibid., citing Gen 1:14.
[131]Ibid. (FC 46:86).
[132]Ibid. (FC 46:88).
[133]Ibid. This is most likely a reference to Paul in Phil 2:15: "In which you shine like stars in the world."

"become lights of the souls whom they have taught, drawing them out from the darkness of ignorance, so also now, having prepared this sun for that most bright light, the Creator of the universe has lighted it around the world."[134]

Basil then brings the discussion back around to helping his hearers understand how light can be created before the vehicle that carries it. All composite bodies are divided into the "recipient substance and the supervenient quality."[135] For example, whiteness by nature is one thing, while a whitened body is something else. Thus, even though the separation of these things is not possible for us, certainly "that which we are able to separate in thought can also be separated in actuality by the Creator of its nature."[136] This is what happened in the incident with the burning bush, "which was active only in its brilliance and had its power of burning inactive."[137] Further, this is what the psalmist indicates when he speaks of "the voice of the Lord dividing the flame of fire."[138]

It is at this point in the second homily that Basil comments specifically on day and night. Genesis 1:5 states, "And there was evening and morning, one day." Here his previous emphasis on the significance of light finds an application. Evening is mentioned first because it is the common boundary line of day and night, while morning is the part of night bordering day. This indicates the "prerogative of prior generation to the day."[139] Since darkness, not night, was the condition of the world before the creation of light, that which was opposed to day was called night and received its name after day. Thus "there was evening and morning" means the space of a day and a night. Now, since day is prior to night and "the important one," the name *day*

---

[134]Ibid.
[135]Ibid., 6.3 (FC 46:86).
[136]Ibid.
[137]Ibid. (FC 46:87).
[138]Ibid., referring to Ps 29:7.
[139]Ibid., 2.8 (FC 46:33).

means day and night.[140] This is a common practice in Scripture.[141] In this way "the words now handed down in the form of history are the laws laid down for later usage."[142]

When Basil gets to the last phrase in Genesis 1:5, "And there was evening and morning, one day," he wonders why the writer used *one* rather than *first*. Some background context is required to understand why Basil answers this question in the way he does – namely, the relationship between sabbath, the Lord's Day, and the eighth day.

The Jewish sabbath stems from Genesis 2:1-2 and Exodus 20:8.[143] In the Jewish calendar it was celebrated on the seventh day of the week, the day which we call Saturday. Early Christians came to lay emphasis on the first day of the week (what we call Sunday) for a few different reasons:[144] it was the day of Christ's post-resurrection appearance to his disciples,[145] it was the day the gift of the Spirit was bestowed,[146] it was the day in which Christ sent his Spirit-enlivened disciples forth as ministers of salvation,[147] and, above all, it was the day of Jesus' resurrection.[148] For Christians this precipitated a new day of worship, the first day of the week, which came to be called the Lord's Day.[149]

The establishment of this new day of worship caused early Christians to consider the place of the Jewish sabbath in their own worship theology. What emerged was a metaphorical understanding of the sabbath, which allowed them "to explain how the commandment [to

---

[140]Ibid. (FC 46:33-34).

[141]Here Basil quotes Ps 90:10: "The days of our years"; Gen 47:9 (LXX): "The days of my life are few and evil"; and Ps 23:6: "All the days of my life."

[142]Basil, *Hexaemeron* 2.8 (FC 46:34).

[143]Gen 2:1-2: "Thus the heavens and the earth were finished, and all their multitude. And on the seventh day God finished the work that he had done, and he rested on the seventh day from all his work that he had done." Ex 20:8: "Remember the sabbath day, and keep it holy."

[144]Thomas K. Carroll and Thomas Halton, *Liturgical Practice in the Fathers*, Message of the Fathers 21 (Wilmington, DE: Michael Glazier, 1988), 19.

[145]Mt 28:1-9; Lk 24:1-12; Jn 20:19-23.

[146]Acts 2:1-4; Jn 20:19-22.

[147]Jn 20:21-23; Acts 1:8.

[148]Mt 28:1-10; Mk 16:1-8; Lk 24:1-12; Jn 20:1-10.

[149]1 Cor 16:2.

honor the sabbath] could be both God-given and valuable, and yet not binding on Christians in its literal sense."[150] This move is well illustrated in the second-century Epistle of Barnabas:

Further, also, it is written concerning the Sabbath in the Decalogue which [the Lord] spoke, face to face, to Moses on Mount Sinai, "And sanctify ye the Sabbath of the Lord with clean hands and a pure heart" [Ex 20:8; Deut 5:12]. And He says in another place, "If my sons keep the Sabbath, then will I cause my mercy to rest upon them" [Jer 17:24]. The Sabbath is mentioned at the beginning of the creation [thus]: "And God made in six days the works of His hands, and made an end on the seventh day, and rested on it, and sanctified it" [Gen 2:2]. Attend, my children, to the meaning of this expression, "He finished in six days." This implieth that the Lord will finish all things in six thousand years, for a day is with Him a thousand years. And He Himself testifieth, saying, "Behold, to-day will be as a thousand years" [Ps 90:4; 2 Pet 3:8]. Therefore, my children, in six days, that is, in six thousand years, all things will be finished. "And He rested on the seventh day." This meaneth: when His Son, coming [again], shall destroy the time of the wicked man, and judge the ungodly, and change the sun, and the moon, and the stars, then shall He truly rest on the seventh day. Moreover, He says, "Thou shalt sanctify it with pure hands and a pure heart." If, therefore, any one can now sanctify the day which God hath sanctified, except he is pure in heart in all things, we are deceived. Behold, therefore: certainly then one properly resting sanctifies it, when we ourselves, having received the promise, wickedness no longer existing, and all things having been made new by the Lord, shall be able to work righteousness. Then we shall be able to sanctify it, having been first sanctified ourselves. Further, He says to them, "Your new moons and your Sabbaths I cannot endure" [Is 1:13]. Ye perceive how He speaks: Your present Sabbaths are not acceptable to Me, but that is which I have made, [namely this,] when, giving rest to all things, I shall make a beginning of the eighth day, that is, a beginning of another world.[151]

Christians sought to clarify elements of continuity and discontinuity between themselves and Judaism. This is why, for the early church

---

[150]Carroll and Halton, *Liturgical Practice*, 24.
[151]*Epistle of Barnabas* 15 (ANF 1:146-47).

father Tertullian, the observance of the sabbath was "temporary."[152] The Scriptures, Tertullian asserts, "point to a Sabbath eternal and a Sabbath temporal."[153] The temporal is human whereas the eternal is divine. He cites Isaiah 66:23 as being fulfilled in Christ when "'all flesh' – that is, every nation – 'came to adore in Jerusalem' God the Father through Jesus Christ His Son as predicted by the prophet."[154] For Tertullian, this means that an eternal sabbath was foretold, making the present one temporary and spiritual. In this context, the first day of the week, Sunday, became a new temporary sabbath that anticipated an eternal sabbath.[155]

The metaphorical approach indicated by Barnabas and Tertullian was already common among Fathers of the late second century. Later writers developed this and started to see "the first day of the old Jewish week as the beginning in time of the eschatological Sabbath of Genesis, a new day of creation, which had in fact begun with the resurrection of Christ on the first day of the Jewish week."[156] This connection of creation on day one to the Lord's Day, the first day of the week, is seen in a homily attributed to Athanasius:

> The Sabbath was therefore the end [*telos*] of the former creation, but the Lord's Day is the beginning [*archē*] of the second creation, in which the Lord renewed and refreshed the archaic creation. So just as he ordered that the original Sabbath day be kept as the memorial of former things, so we honor [now] the Lord's Day as a commemoration of the second creation. For he did not create another creation but renewed the original one and finished what he had begun to make. . . . For the work was incomplete if, when Adam sinned, humanity had simply died out; but instead it was perfected when he was made alive again. For this reason, renewing the creation that he had made in six days, the Lord appointed a day for its recreation, which the Spirit announced

---

[152]Tertullian, *An Answer to the Jews* 4 (ANF 3:155).
[153]Ibid.
[154]Tertullian cites Is 66:23 from the LXX: "And there shall be month after month, and day after day, and sabbath after sabbath; and all flesh shall come to adore in Jerusalem, saith the Lord."
[155]Carroll and Halton, *Liturgical Practice*, 24.
[156]Ibid., 29.

beforehand through the Psalm, "This is the day that the Lord has made" (Ps 117 [118]:24).[157]

This brings us to the so-called eighth day, which grows out of understandings already expressed.[158] In Judaism, seven and eight were sacred numbers.[159] The number seven was sanctified by the Creator at creation, and the number eight signified redemption and marked the covenant.[160] The mid-second-century apologist Justin Martyr exploits this connection of covenant, circumcision, and the number eight:

> The command of circumcision, again, bidding [them] always circumcise the children on the eighth day, was a type of the true circumcision, by which we are circumcised from deceit and iniquity through Him who rose from the dead on the first day after the Sabbath, [namely through] our Lord Jesus Christ. For the first day after the Sabbath, remaining the first of all the days, is called, however, the eighth, according to the number of all the days of the cycle, and [yet] remains the first.[161]

Justin indicates both a new beginning as well as a renewal of a cycle – the new beginning is the first day of the week on which Christ rose from the dead, but since it recurs and renews itself in keeping with the cycle of the week, it is also called the eighth. So it is that from the fourth century onward, differentiation between the Jewish sabbath and the Christian Lord's Day led to "claims of the eschatological superiority of Sunday, the Day of Resurrection, as a transcendent 'Eighth Day,' the day of the spiritual circumcision, and so too a whole new beginning for God's creation under the Christian dispensation."[162]

---

[157]Athanasius, *On the Sabbath and Circumcision* 4-5, cited from Paul Blowers, *Drama of the Divine Economy: Creator and Creation in Early Christian Theology and Piety*, Oxford Early Christian Studies (Oxford: Oxford University Press, 2012), 340.

[158]What follows draws from Carroll and Halton, *Liturgical Practice*, 35-36.

[159]Eccles 11:2: "Divide your means seven ways, or even eight, for you do not know what disaster may happen on earth."

[160]Gen 17:11-12: "You shall circumcise the flesh of your foreskins, and it shall be a sign of the covenant between me and you. Throughout your generations every male among you shall be circumcised when he is eight days old."

[161]Justin Martyr, *Dialogue with Trypho* 41 (ANF 1:215).

[162]Blowers, *Drama of the Divine Economy*, 339.

In the church of Basil's time, the Easter Octave (the Sunday after Easter) was celebrated. It was so named (octave = eight) because the day on which it was celebrated was eight days after Easter (Easter plus the intervening six days). Because this was a festival of renewal it shows how Sunday could be identified as both eighth day and creation day.[163] Basil's contemporary and close friend, Gregory of Nazianzus, is particularly helpful here:

> "What is the point?" someone may ask. "Was not the First Sunday [Pascha] the feast of our renewal – the day after that holy night made bright by our candles? Why are you proclaiming it today? Are you simply a lover of festivals, inventing a multiplicity of splendid occasions?" Last Sunday was the day of salvation, but today is salvation's anniversary. Last Sunday revealed the boundary between the grave and the resurrection, but today reveals, in all its clarity, our second becoming. *So just as the first creation has its beginning on a Sunday (and that is clear: for the seventh day after it is the Sabbath, which brings cessation from work), so the new creation must begin again with the same day: the first of the days that come after it, the eighth of those that come before, more exalted than what has been exalted before, more wonderful than previous wonders.* For it refers to the state of things that lies beyond us, which holy Solomon seems to hint at when he commands that we "give a portion to seven" – that is, to this life – "or even to Eight" (Eccles 11:2) – that is, to the life to come – on the basis of our good works in this life and of the restoration of all things in the next.[164]

According to Paul Blowers there existed an octave structure (first day → six days → eighth day) that became important in the liturgy and spiritual formation of Christians.[165] In a homily on Psalm 6, a psalm that captured the attention of the Fathers because of its title "To the Eighth," Basil's brother, Gregory of Nyssa, also shows the strong connection between the first day (creation) and the eighth day (new creation):

---

[163]Ibid., 341.
[164]Gregory of Nazianzus, *Oration* 44, cited in Blowers, *Drama of the Divine Economy*, 341-42. Italics added.
[165]Blowers, *Drama of the Divine Economy*, 343.

The time in which we live this life, in the first formative period of creation, has reached its completion in a week of seven days. The shaping of all that exists began with the first day, and the outer limit of creation came to its final form on the seventh. "For there was one day" (Gen. 1:5), Scripture says, in which the first stage of things came to be; similarly, too, the second stage took place on the second day, and so onwards in sequence until all the works of the six days had been completed. The seventh day, which came to define in itself the limit of creation, set the boundary for the time that is co-extensive with all the furnishings of the world. As a result, no other heaven has come to be from this one, nor has any other part of the world been added to those that exist from the beginning; creation has come to rest in what it is, remaining complete and undiminished within its own boundaries. So, too, no other time has come into existence alongside that time that was revealed along with the formation of the world, but the nature of time has rather been circumscribed by the week of seven days. For this reason, when we measure time in days, we begin with day one and close the number with day seven, returning then to the first day of the new week; so we continue to measure the whole extension [diastēma] of time by the cycle of weeks, until – when the things that are in motion pass away and the flux of the world's movement comes to an end – "those things" come to be, as the apostle says, "which will never be shaken" (Heb. 12:27), things that change and alteration will never touch. This [new] creation will always remain unchanged, for the ages to follow; in it, the true circumcision of human nature will come to reality.[166]

We see all the elements we have been discussing thus far come together in Basil's famous book *On the Holy Spirit*. This prepares us to understand the claims he makes in the second homily on the Hexaemeron. Making specific reference to the liturgical practice of standing during prayers, Basil states:

> We pray standing, on the first day of the week, but we do not all know the reason. On the day of the resurrection [or "standing again," Greek *anastasis*] we remind ourselves of the grace given to us by standing at prayer, not only because we rose with Christ, and are bound to "seek those things which are above," but because the day seems to us to be in

---

[166]Gregory of Nyssa, *Homily on Psalm 6*, cited in Blowers, *Drama of the Divine Economy*, 343.

some sense an image of the age which we expect, wherefore, though it is the beginning of days, it is not called by Moses first, but one. For he says "There was evening, and there was morning, one day," as though the same day often recurred. Now "one" and "eighth" are the same, in itself distinctly indicating that really "one" and "eighth" of which the Psalmist makes mention in certain titles of the Psalms,[167] the state which follows after this present time, the day which knows no waning or eventide, and no successor, that age which endeth not or groweth old. Of necessity, then, the church teaches her own foster children to offer their prayers on that day standing, to the end that through continual reminder of the endless life we may not neglect to make provision for our removal thither. Moreover all Pentecost is a reminder of the resurrection expected in the age to come. For that one and first day, if seven times multiplied by seven, completes the seven weeks of the holy Pentecost; for, beginning at the first, Pentecost ends with the same, making fifty revolutions through the like intervening days. And so it is a likeness of eternity, beginning as it does and ending, as in a circling course, at the same point. On this day the rules of the church have educated us to prefer the upright attitude of prayer, for by their plain reminder they, as it were, make our mind to dwell no longer in the present but in the future. Moreover every time we fall upon our knees and rise from off them we shew by the very deed that by our sin we fell down to earth, and by the loving kindness of our Creator were called back to heaven.[168]

Here we see Basil connecting the "first day of the week" with the "day of the resurrection," which was also called the Lord's Day. In the worship of the church during Basil's time these were the same day, what we call Sunday, which was emphasized as the *archē* (principal, beginning) of the week.[169] The passage in *On the Holy Spirit* deals with explaining the symbolism of the rites in the church to the faithful.[170]

---

[167]The title of Ps 6 and Ps 11 is "Concerning the Eighth."

[168]Basil of Caesarea, *On the Holy Spirit* 27.66 (NPNF² 8:42-43).

[169]Carroll and Halton, *Liturgical Practice in the Fathers*, 10.

[170]Jean Daniélou, *The Bible and the Liturgy*, Liturgical Studies 3 (Notre Dame, IN: University of Notre Dame Press, 1961), 263. This explanation is called "mystagogy" and can be described as a common teaching practice of the early church, particularly that of the fourth and fifth centuries. It was the teaching on the meaning of the church rites that were usually given by a bishop to those undertaking pre-baptismal catechetical training.

Standing to say prayers, he explains, is meant to be a reminder of grace given on the day of the resurrection – "as if we are rising and standing with Christ and being bound to seek what is above."[171] Since the Lord's Day is a remembrance of the resurrection, it is "an image [*eikon*] of the age to come." This points to its eschatological significance wherein the Christian is called to not become entangled in the things of the earth but to "seek what is above."[172] Here we have what Daniélou calls a "theology of the Lord's Day" that is formed in the image "of Christian worship in general, in which the double aspect of memorial and of prophecy is clearly marked."[173]

It is striking that Basil connects the day of the resurrection to Genesis 1:5 and the beginning of days in this second homily. Even though Moses refers to the beginning of days, he calls it "one" and not "first." We now have the context within which we can understand Basil's comment about "one" and "eighth" being the same: "the really 'one' and the 'eighth' . . . are the state after this time, the unceasing, unending, perpetual day, that never-ending and ever-young age."

Basil develops this further in our passage (*Hexaemeron* 2.8). To him, someone trying to introduce a second and a third and a fourth day would be more consistent to call the beginning of the series "first" rather than "one." He gives several possible answers to the query. First, God may have wanted to determine precisely the duration of a full day as a twenty-four-hour unit of time.[174] "He said 'one' because he was defining the measure of day and night and combining the time of a

---

[171]Cf. Col 3:1.

[172]In typical theological language, eschatology is that branch of theology dealing with last things or end times. In the church fathers, eschatology is not simply the era of earthly time followed by the era of the eternal. It is, rather, a "piercing" of the eternal into the temporal, or perhaps an even better analogy would be the "infection" of the temporal with the eternal (cf. C. S. Lewis, *Mere Christianity* [San Francisco: HarperSanFrancisco, 2001], book 4, chapter 4, titled "Good Infection"). My thanks to Hanna Lucas for this helpful reference to Lewis and the explanation in her Trinity Western University master's thesis ("To Extend the Sight of the Soul: An Analysis of Sacramental Ontology in the Mystagogical Homilies of Theodore of Mopsuestia," 33).

[173]Daniélou, *Bible and the Liturgy*, 264.

[174]Callahan, "Greek Philosophy," 37.

night and day, since the twenty-four hours fill up the interval of one day, if, of course, night is understood with day." This, says Basil, is as if one was saying that twenty-four hours is the measure of one day. For Basil, this indicates that the revolution of the sun around the earth completes a circle in the space of one day. Basil's second answer to the question of Moses' use of "one" rather than "first" is related – that each time there is a revolution of the heavens this will be considered one day, as a unit for reckoning time.[175]

For Basil, time is "a kind of receptacle which . . . encloses the evolution of the world from beginning to end. It . . . envelops the motion of the created order."[176] Time is not eternal; it has a beginning and an end and will finish with the universe.[177] Days, weeks, months, and years are not elements or parts of time, they merely measure and signify its passage and motion.[178] This is why the sun and the moon do not make day and night, but only mark them. Light and darkness existed before the creation of the sun and moon, so their movement only signifies the separation of day and night – they do not cause them.

This is why Basil moves on to his third answer, which relates the days to eternity:

> Or is the reason [for "one day" rather than "first day"] handed down in the mysteries more authoritative, that God, having prepared the nature of time, set as measures and limits for it the intervals of the days, and measuring it out for a week, He orders the week, in counting the change of time, always to return again in a circle to itself? Again, He orders that one day by recurring seven times complete a week; and this, beginning from itself and ending on itself, is the form of a circle. In fact, it is also characteristic of eternity to turn back upon itself and never to be brought to an end. Therefore, He called the beginning of time not a "first day" but "one day" in order that from the name it might have kinship with eternity.

---

[175]Ibid., 38.
[176]Orphanos, *Creation and Salvation*, 53-54. Cf. Basil, *Hexaemeron* 1.5; and Basil, *Against Eunomius* 1.21.
[177]Basil, *Hexaemeron* 1.5 (FC 46:9).
[178]Basil, *Against Eunomius* 1.21.

> For, the day which shows a character of uniqueness and nonparticipation
> with the rest is properly and naturally called "one."[179]

The reference to that which was "handed down in the mysteries" refers
to the liturgical context we saw in *On the Holy Spirit*, wherein the rites of
the church are explained to the faithful. The idea we saw above in *On
the Holy Spirit* is here given more detail – that is, "if the day which begins
the week is called 'one' it is to indicate that the week, returning on
itself, forms a unity."[180] Two ideas should be highlighted here: that the
world of time is ruled by a seven-day period, and that the week represents a closed cycle returning perpetually on itself, having neither beginning nor end and representing eternity.[181] Basil is advocating for the
connaturality between the one and the seven. He continues in this vein:

> If, however, the Scripture presents to us many ages, saying in various
> places "age of age" and "ages of ages" [Ps 23:6; 148:6; Jude 25] still in those
> places neither the first, nor the second, nor the third age is enumerated
> for us, so that, by this, differences of conditions and of various circum-
> stances are shown to us but not limits and boundaries and successions
> of ages.[182]

According to Daniélou, the ages here described are not ages in suc-
cessive time but rather "universes differing in quality." This means
that each age is properly individual and cannot be counted among
others. Each age is thus "one," but not first or second.[183] Cyclical time,
"beginning from itself and ending on itself," is the image of this kind
of unity.[184]

Basil clarifies the connection between these speculations about time
and the worship of the church:

---

[179]Basil, *Hexaemeron* 2.8 (FC 46:34-35). Italics in original translation.

[180]Daniélou, *Bible and the Liturgy*, 265. Daniélou here cites the sixth-century writer John Lydus's
De mensibus to show that this is a Pythagorean idea: "The first day according to the Pythago-
reans, ought to be called one (*mia*) of the monad [one], not the first (*prōtē*) of the hebdomad
[seven], because it is unique and cannot be communicated to others."

[181]Daniélou, *Bible and the Liturgy*, 265; and Carroll and Halton, *Liturgical Practice*, 11.

[182]Basil, *Hexaemeron* 2.8 (FC 46:35).

[183]Daniélou, *Bible and the Liturgy*, 265.

[184]Basil, *Hexaemeron* 2.8 (FC 46:34).

"The day of the Lord is great and very terrible" [Joel 2:11] it is said. And
again, "To what end do you seek the day of the Lord? And this is darkness,
and not light" [cf. Amos 5:18]. But darkness, certainly, for those who are
deserving of darkness. For, Scripture knows as a day without evening,
without succession, and without end, that day which the psalmist called
the eighth, because it lies outside this week of time. Therefore, whether
you say "day" or "age" you will express the same idea. If, then, that condition should be called day, it is one and not many, or, if it should be
named age, it would be unique and not manifold. In order, therefore, to
lead our thoughts to a future life, he called that day "one" which is an
image of eternity, the beginning of days, the contemporary of light, the
holy Lord's day, the day honored by the Resurrection of the Lord. "There
was, then, evening and morning, one day" he said.[185]

The Lord's Day is here connected with the eighth day, which is
"outside this week of time." This day is "without evening, without succession, and without end." These are the same expressions used in the
passage *On the Holy Spirit* above. The first day of the week, in Basil's
thinking, serves as the visible symbol that leads toward eternity. The day
on which light was created and the day on which Jesus rose from the
dead is commemorated every week in the Sunday worship of the church.
There is unity in each of these days or ages, and this then signifies the
oneness of the age to come. Thus, "the whole theology of the Sunday is
now seen clearly; it is the cosmic day of creation, the biblical day of
circumcision, the evangelical day of the Resurrection, the Church's day
of the Eucharistic celebration, and, finally, the eschatological day of the
age to come."[186]

Basil is making a fundamental contrast between the seven days, which
is a figure of the present world, and the eighth day, which is a figure of
the world to come. The Lord's Day is both "one" and "the eighth." It is
"one" in that the future life is "one" without any succession of time. It is
"eighth" in that "seven days" is a figure of the world to come.

---

[185]Ibid. (FC 46:35).
[186]Daniélou, *Bible and the Liturgy*, 266.

Basil's conclusion to this second homily seems to betray the intended emphasis of his words:

> But, in truth, my words concerning that evening, being overtaken by the present evening, mark the end of my speech. May the Father of the true light, however, who has decked the day with the heavenly light, who has brightened the night with gleams of fire, who has made ready the peace of the future age with a spiritual and never ending light, illumine your hearts in a knowledge of the truth, and preserve your life without offense, allowing you "to walk becomingly as in the day" [cf. Rom 13:13] in order that you may shine forth as the sun in the splendor of the saints for my exultation in the day of Christ, to whom be glory and power forever. Amen.[187]

RECOMMENDED READING

Blowers, Paul. *Drama of the Divine Economy: Creator and Creation in Early Christian Theology and Piety*. Oxford Early Christian Studies. Oxford: Oxford University Press, 2012.

Bouteneff, Peter C. *Beginnings: Ancient Christian Readings of the Biblical Creation Narratives*. Grand Rapids: Baker Academic, 2008.

Louth, Andrew. "The Six Days of Creation According to the Greek Fathers." In *Reading Genesis After Darwin*, edited by Stephen C. Barton and David Wilkinson, 39-55. Oxford: Oxford University Press, 2009.

---

[187]Basil, *Hexaemeron* 2.8 (FC 46:35-36).

7

AUGUSTINE *on* "IN *the* BEGINNING"

Let's start at the very beginning, a very good place to start.

To my knowledge, no one has really taken issue with the opening lyrics to Rodgers and Hammerstein's well-known song from *The Sound of Music*. When we hear Julie Andrews's voice sing, "Let's start at the very beginning / A very good place to start," the words appear to us as common sense. But some reflection on the word and concept allows us to see that, understood in the sense of a point of departure, *beginning* actually has different shades of meaning.[1] It can be connected to time, as when, during a long flight from London to Vancouver, Captain Sorenson comes on the PA and announces, "Ladies and gentlemen, we are beginning our descent into Vancouver." But a beginning or a point of departure could also be a reference to a basis or a rationale, a purpose, or a reason — that is, something that is tied more to a source of origin. For example, I might say about my neighbor, "Rick's shoddy lawn was the basis (or beginning) of my love for proper lawn care," or a scientist

---

[1]Examples adapted from R. R. Reno, *Genesis*, Brazos Theological Commentary on the Bible (Grand Rapids: Brazos Press, 2010), 30.

might say, "The second law of thermodynamics is the beginning (basis) of cosmology."

The difference here is between a temporal claim (the pilot) and a substantive claim (the neighbor and scientist). When we read in Psalm 111:10 that "The fear of the LORD is the beginning of wisdom," we read it as a substantive claim rather than a temporal claim. That is, we understand this verse as indicating that "fear of the Lord is the origin of the wise life, not in the sense of the first step that is superseded by the second and third, but in the lasting sense of providing its basis or root."[2]

Awareness of the well-accepted difference between a temporal and a substantive claim places us in a good position to begin to understand how Augustine reads the first words of Genesis. These words do not introduce mere reporting of facts or events for Augustine. Rather, they have a significance that is ultimately rooted in Christ. In order to understand the conclusion Augustine makes in this regard, we need to cover some contextual ground in his writings.

WHAT WAS GOD DOING BEFORE CREATION?

The wide Christian acceptance of *creatio ex nihilo* by Augustine's time led to a body of well-known questions posed by philosophers who argued against a beginning of the world.[3] Augustine shows his awareness of these questions in all his writings on Genesis.[4] In *The Confessions* he relates a question that some ask of Genesis 1: "What was God doing before he made heaven and earth? . . . If he was at leisure . . . and not making anything, why did he not continue so thereafter and

---

[2]Ibid.

[3]For a detailed discussion of the body of questions, see Edward Peters, "What Was God Doing Before He Created the Heavens and the Earth?," *Augustiniana* 34 (1984): 53-74; and Richard Sorabji, *Time, Creation and the Continuum: Theories in Antiquity and the Early Middle Ages* (Ithaca, NY: Cornell University Press, 1983), 232-52.

[4]Augustine devoted more attention to Genesis than any other early Christian writer: See *On Genesis: A Refutation of the Manichees* (388); *Unfinished Literal Commentary on Genesis* (393-95); *The Confessions*, books 12 and 13 (397-401); *The Literal Meaning of Genesis* (401); and *The City of God*, book 11 (413-27).

forever, just as he had always done nothing prior to that?"[5] This question occupies the entirety of the eleventh book of *The Confessions* and, in fact, serves to represent all wrong questions in this vein.[6]

In his first foray into interpreting Genesis, Augustine showed how the Manichees contributed to this body of questions. The Manichees find fault with the first verse of Genesis – particularly with the phrase "In the beginning." "In what beginning?" they ask. If heaven and earth were made by God in some beginning of time, what was he doing before he created them? Further, "Why did it suddenly take his fancy to make what he had never made previously through eternal times?"[7] In *The Literal Meaning of Genesis*, Augustine frames this question from a more positive perspective: "And how could it be shown that God produced changeable and time-bound works without any change in himself?"[8]

In *The City of God*, Augustine again shows his awareness of this body of commonly posed questions: "But why did the eternal God decide to make heaven and earth just then, when he had not made them before?"[9] He takes issue with anyone who would claim that the world is eternal, without any beginning, and therefore not created by God. This claim is made by those who "have gone far astray from the truth." Even without Holy Scripture, Augustine claims, the order, movement, and beauty of visible things "silently proclaims both that it was made and that it could only have been made by God, who is himself inexpressibly and invisibly great and inexpressibly and invisibly beautiful."[10]

---

[5]Augustine, *Confessions* 11.10.12 (TWSA I/1:230). All quotations from *The Confessions* are from Saint Augustine, *The Confessions*, trans. Maria Boulding, TWSA I/1 (Hyde Park, NY: New City Press, 2001).
[6]Peters, "What Was God Doing?," 54-55.
[7]Augustine, *On Genesis: A Refutation of the Manichees* 1.2.3 (TWSA I/13:40). All quotations of this work are from Saint Augustine, *On Genesis: A Refutation of the Manichees*, trans. Edmund Hill, TWSA I/13 (Hyde Park, NY: New City Press, 2002).
[8]Augustine, *The Literal Meaning of Genesis* 1.2 (TWSA I/13:168). All quotations of *The Literal Meaning of Genesis* are from Saint Augustine, *The Literal Meaning of Genesis*, trans. Edmund Hill, TWSA I/13 (Hyde Park, NY: New City Press, 2001).
[9]Augustine, *The City of God* 11.4 (TWSA I/7:4). All quotations from *The City of God* are from Saint Augustine, *The City of God*, books 11–22, trans. William Babcock, TWSA I/7 (Hyde Park, NY: New City Press, 2013).
[10]Ibid., 11.4 (TWSA I/7:4).

Some meet the query, "What was God doing?" with the quip "He was getting hell ready for people who inquisitively peer into deep matters." But Augustine is not satisfied to dismiss it so easily, because it evades the force of the question.[11] He considers it an important question to address because it stems from a failure to understand the limitations of the concept of time as it applies to the eternal God and Word. Some philosophers considered creation as implying an act of will at some point in time by God, and an act of will implies change in God: "If some change took place in God, and some new volition emerged to inaugurate created being, a thing he had never done before, then an act of will was arising in him which had not previously been present, and in that case how would he truly be eternal?"[12]

TIME AND ETERNITY

Augustine argues that this body of questions stems from a failure to understand the distinction between time and eternity. In book 11 of *The Confessions* he addresses the meaning of *beginning* in Genesis 1:1 through a discussion of time and eternity. Eternity, which belongs only to God, is distinguished from time. God is not conditioned by time whereas everything else is.[13] Here the context of praise to the eternal God frames Augustine's prayer for understanding the Scriptures, and he moves from a prayer of praise to a more extended discussion of *beginning* in Genesis 1:1 on the foundation of this distinction.

The distinction between eternity and time is furthered through an explanation of change and variation, which are indicators of creation. When things undergo change and variation it indicates that they were created. "Heaven and earth," Augustine states, "proclaim that they did not make themselves."[14] Of course Augustine believes it was God who made them, and because he made them they are beautiful, good, and

---

[11]Augustine, *Confessions* 11.12.14 (TWSA I/1:231).
[12]Ibid., 11.10.12 (TWSA I/1:230).
[13]Ibid., 11.1.1 (TWSA I/1:223).
[14]Ibid., 11.4.6 (TWSA I/1:226).

have being. But the beauty, goodness, and being of God's creation are not so in the same way that God is beautiful, good, and has being. Nothing exists as God exists.

The questions of the philosophers stem also from a failure to realize that all things come into existence by God and have their being in him. Augustine explains, "They strive to be wise about eternal realities, but their heart flutters about between the changes of past and future found in created things, and an empty heart it remains."[15] There is no time in eternity, and all of both past and future comes from God, who is always present and who created time.[16] To imagine that God almighty, creator of all things, was at rest throughout the measureless ages before creation is ridiculous: "How could measureless ages have passed by if you [God] had not made them, since you are the author, and creator of the ages? Or what epochs of time could have existed, that had not been created by you? And how could they have passed by, if they had never existed?"[17] There was no time before creation, so to ask what God was doing before creation domesticates his eternality in the change and flux of time, which is itself a creation.[18]

Even to claim that God is "earlier in time" than all eras is inaccurate because this would imply that time was in existence before God. God's "ever-present eternity" means that he is neither governed nor bound by

---

[15]Ibid., 11.11.13 (TWSA I/1:230).

[16]Augustine, *On Genesis* 1.2.3 (TWSA I/13:41): "Why did it suddenly take his fancy? The question assumes that a lot of time had passed in which God had not constructed anything. But after all, no time, which God had not yet made, could have passed, since the constructor of times must be there before the times."

[17]Augustine, *Confessions* 11.13.15 (TWSA I/1:231).

[18]Cf. Augustine, *On Genesis* 1.2.4 (TWSA I/13:41): "So then, if these people ever say, 'Why did it take God's fancy to make heaven and earth?' the answer to be given is that those who desire to know God's will should first set about learning the force of the human will. You see, they are seeking to know the causes of everything there is. After all, if God's will has a cause, there is something that is there before God's will and takes precedence over it, which it is impious to believe. So then, anyone who says, 'Why did God make heaven and earth?' is to be given this answer: 'Because he wished to.' It is God's will, you see, that is the cause of heaven and earth, and that is why God's will is greater than heaven and earth. Anyone, though, who goes on to say, 'Why did he wish to make heaven and earth?' is looking for something greater than God's will is; but nothing greater can be found. So let mere human beings put a curb on their brashness, and not go seeking what doesn't exist, in case they thereby fail to find what does."

time. "Your today is eternity, and therefore your Son, to whom you said, *Today have I begotten you* [Ps 2:7; Act 13:33; Heb 1:5; 5:5], is coeternal with you. You have made all eras of time and you are before all time, and there was never a 'time' when time did not exist."[19]

This leads Augustine into his famous excursus on time in *The Confessions*. Proceeding on the foundation already established that time is created by God, Augustine answers the question, "What is time?" We speak about time with great familiarity and ease. It is a common experience shared by all. But if we stop to ponder the concept of time we notice its evasiveness. Time, he concludes, does not actually exist:

> Now, what about those two times, past and future: in what sense do they have real being, if the past no longer exists and the future does not yet exist? As for the present time, if that were always present and never slipped away into the past, it would not be time at all; it would be eternity. If, therefore, the present's only claim to be called "time" is that it is slipping away into the past, how can we assert that this thing *is*, when its only title to being is that it will soon cease to be? In other words, we cannot really say that time exists, except because it tends to non-being.[20]

Time, for Augustine, is a measure of change; it is a relation among changing things. Because of this, it exists only where change exists. As William A. Christian explains, "Thus, the original come-into-being of the world of changing creatures 'begins' the temporal process and does not fall within that process."[21] Further, because God is eternal, and creation is an act of God's will, his willing is also eternal – it is not in time, nor does it take time.

Augustine concludes book 11 of *The Confessions*, and his distinction between time and eternity, in the same way he began – by offering a prayer focusing on God's eternality.[22] Eternity is stable (in it there is

---

[19]Augustine, *Confessions* 11.13.16 (TWSA I/1:232). Italics in original translation.

[20]Ibid., 11.14.17 (TWSA I/1:232-33).

[21]William A. Christian, "Augustine and the Creation of the World," *Harvard Theological Review* 46, no. 1 (1953): 4.

[22]Augustine, *Confessions* 11.29.39 (TWSA I/1:245-47).

no change) whereas creaturely existence implies change and flux because it is in time. We now see that Augustine's discussion of time is bookended by the question, "What was God doing before he made heaven and earth?"[23] Here he refers to those who ask a related question: "Why did it enter his head to make something, when he had never made anything before?" His answer is:

> Grant them, Lord, the grace to think clearly what they are saying, and to realize that the word "never" has no meaning where time does not exist. If God is said never to have done something, that simply means that he did not do it at any time. Let such people see then, that there cannot be any time apart from creation, and stop talking about nonsense. Let them even stretch their minds to what lies ahead and understand that you exist before all ages of time, because you are the eternal creator of all times, and that no time is coeternal with you, nor any creature whatsoever, even if any was created outside of time.[24]

> Nothing can happen to you in your unchangeable eternity, you who are truly the eternal creator of all minds. As you knew heaven and earth in the beginning, without the slightest modification in your knowledge, so too you made heaven and earth in the beginning without any distension in your activity.[25]

Augustine addresses the body of questions more specifically in *The City of God*. He takes issue with those who admit that God did, indeed, create the world, but say that it has no beginning in time, "only a beginning in the sense of creation," so that in a confusing way, it was always created.[26] This appears to defend God against the charge of acting on random impulse, as if it suddenly occurred to God to create the world, as if it had never occurred to him before, or that he had a change of will. But the truth is that God is immutable – the world was created in time, yet in making it God's eternal will and design were in no way changed.

---

[23]Stated in ibid., 11.10.12 (TWSA I/1:230) and again in 11.30.40 (TWSA I/1:246).
[24]Ibid., 11.30.40 (TWSA I/1:246).
[25]Ibid., 11.31.41 (TWSA I/1:247).
[26]Augustine, *City of God* 11.4 (TWSA I/7:4).

Ultimately Augustine's answers to these questions, again, revolve around the proper distinction between time and eternity.[27] Time does not exist without some movement and change, but there is no movement or change in eternity. Time simply would not exist if no creature had been made to bring about change by means of some motion.

> Therefore, since God, in whose eternity there is no change whatsoever, is the creator and governor of time, I do not see how it can be said that he created the world after expanses of time, unless it is claimed that, prior to the world, there was already some created being by virtue of whose motions time was able to pass.[28]

For Augustine, the "holy and utterly truthful Scriptures" tell us that God made heaven and earth in the beginning so that we might know that nothing was made prior to this. If this were not so, he says, Scripture would have told us. Thus, it is "beyond doubt" that the world was not created *in* time but rather *with* time. No time could have passed when there was no created being, because created beings provide the change and motion of which time is a function. Therefore, the world was made with time since change and movement were created when the world was created; there was no "time" before creation. Susan Schreiner calls this Augustine's "metaphysic of Being or Essence."[29] God did not precede creation by time but rather by an eternity that is above the temporal process.

TIME, ETERNITY, AND THE DAYS OF GENESIS

The understanding that the world was created *with* time rather than *in* time appears to be represented by the order of the "first six or seven days." Both the morning and the evening of these days are described right up until the sixth day, when all that is created by God is completed

---

[27]Ibid., 11.6 (TWSA I/7:6-7).

[28]Ibid., 11.6 (TWSA I/7:7).

[29]Susan E. Schreiner, "Eve, the Mother of History: Reaching for the Reality of History in Augustine's Later Exegesis of Genesis," in *Genesis 1–3 in the History of Exegesis: Intrigue in the Garden*, ed. Gregory Allen Robbins (Lewiston, NY: Edwin Mellen Press, 1988), 157.

and he begins his rest on the seventh. Augustine calls this "God's rest" and states that it "is presented as a great mystery."[30] Yet even in this apparent nod to the temporality of the days in Genesis, Augustine asks us to consider what *days* indicates here: "What kind of days these are it is extremely difficult, or even impossible, to conceive, let alone put into words."[31]

With this self-imposed caution, Augustine proceeds to explain why it is difficult, if not impossible, to understand what kind of days are referenced in Genesis 1. He famously points out that the sun was made on the fourth day, while the first three days passed by without the sun (Gen 1:14-19).[32] If the days known to us have evenings and mornings only because of the rising and setting of the sun, how can this be? Genesis 1:3-4 tells us that, at first, God made the light by his word and that he separated the light from the darkness. The light was called day and the darkness night. But what kind of light it was, and by what alternating motion it made evening and morning, and what sort of evening or morning these were "are all things removed from our senses." Either this was a corporeal light or the word *light* was used to indicate the holy city, made up of angels and blessed spirits, to which Paul refers in Galatians 4:26 and 1 Thessalonians 5:5. But, he continues, if this is a reference to the holy city of spiritual beings, then we need to figure out what the evening and the morning of this day are.

To do this Augustine compares the six days of creation analogically with the creature's knowledge. The creature's knowledge is like "the dusk of evening," but when it is directed in praise and love to the Creator, "it becomes the full light of morning."[33] When Genesis 1 lists the first days in order, Augustine points out that it never uses the word *night* but instead states that "evening came and morning came, one day" (Gen 1:5). The same is true for the rest of the evenings and days.

---

[30]Augustine, *City of God* 11.6 (TWSA I/7:7).
[31]Ibid., 11.6 (TWSA I/7:7).
[32]Ibid., 11.7 (TWSA I/7:7).
[33]Ibid. (TWSA I/7:8).

Knowledge of a creature, taken in itself, pales in comparison with when it is known in the wisdom of God, "in the art by which it is made." This is why it is more appropriate to call it evening rather than night; when it is directed to the Creator in praise and love it turns again into morning.

> And when it does this in its knowledge of itself, this is the first day. When it does this in its knowledge of the firmament, which separates the lower waters from the upper waters and is called the heaven, that is the second day. When it does this in its knowledge of the earth and the sea and all that grows from the earth and has its roots in it, that is the third day. When it does this in its knowledge of the greater and lesser lights of heaven and all the stars, that is the fourth day. When it does this in its knowledge of all the living things that swim in the waters and of all that fly, that is the fifth day. And when it does this in its knowledge of all the beasts of the earth and of man himself, that is the sixth day.[34]

GOD'S REST IN *THE LITERAL MEANING OF GENESIS*

On the seventh day God rests from all his works and sanctifies that day. Augustine explains that just as the first six days are not to be understood in their literal sense, neither is the idea of God resting: "We are not to understand this in a childish way, as if God labored at his work."[35] Rather, God spoke and the heavens and earth were created (Ps 33:9; 148:5). In keeping with Augustine's belief that this word was intelligible and eternal rather than audible, God's rest signifies those who rest in God.

Augustine does not go into any more detail about God's rest and sabbath rest in this section, but he returns to it in the final pages of *The City of God*.[36] Before we turn to that passage we can, however, further understand Augustine's concept of rest as it relates to the six days and the seventh day in Genesis by turning to *The Literal Meaning of Genesis*.[37] Again, just as in *The City of God*, Augustine does not want his readers to

---

[34]Ibid. (TWSA I/7:8).
[35]Ibid., 11.8 (TWSA I/7:8).
[36]Augustine, *City of God* 22.30 (TWSA I/7:551-54).
[37]The extended section with which we will deal is Augustine, *Literal Meaning of Genesis* 4.16–4.20.37 (TWSA I/13:250-62).

think about God resting from his works in a "materialistic" way, as if God needed rest. After all, God "spoke and they were made [Ps 148:5]."[38] How can that require rest? In what sense is God said to rest? Augustine gives us two suggestions.

He first suggests that perhaps God's rest was more emotional and intellectual than physical – like someone who has just completed stressful business dealings. Not much consideration to this is given because it is "the height of folly to entertain" it. The second suggestion about the meaning of God's rest is given much more consideration. Augustine argues here that human creation was granted rest in him – "we rest thanks to his munificence."[39] This understanding is accepted as true because he sees this kind of thing in biblical descriptions of God's knowledge. "God is said to rest when he makes us rest, just as he is said to come to know when he ensures that we come to know something."[40] To Abraham, for example, God says, "Now I have come to know that you fear God."[41] Since God does not come to know anything in time that he did not know beforehand, this is a manner of speech used to mean something like "Now I have made you come to know." Scriptural usage sets the precedent for this type of thing, but, in reality, "we certainly ought not to say anything of the sort about God, which we do not read in his scriptures."[42]

This explanation is not quite enough for Augustine. While it is true that God "makes us rest when we have done good works," he still wants to inquire whether God himself could properly be said to rest.[43] He continues investigating this question by referring to the creation of heaven and earth and all the things in them that were finished on the sixth day. Certainly we did not create any of these things, which means

---

[38]Ibid., 4.16 (TWSA I/13:250).
[39]Ibid., 4.16.9 (TWSA I/13:250).
[40]Ibid., 4.17 (TWSA I/13:251).
[41]Gen 22:12, as cited by Augustine.
[42]Augustine, *Literal Meaning of Genesis* 4.17 (TWSA I/13:251). Augustine proceeds also to give the Pauline examples of Eph 4:30 and Gal 4:9 in 4.18-19.
[43]Augustine, *Literal Meaning of Genesis* 4.10.20 (TWSA I/13:252).

there is a sense in which the biblical claim that "God rested on the seventh day from all his works" is emphatically *not* our rest. After all, the text states that *God* rested. Augustine admits it is quite right to claim that, just as God rested after his good works, we too will rest after our good works. But that is not enough for him, and "so too we should reflect upon God's rest which is shown properly to be his."[44]

So, what is God's rest? Augustine begins investigating this question with John 5:17: "My Father is still working, and I also am working."[45] How is it that God can be said to rest yet also still working? The statement Augustine uses from the Gospel of John is the answer Jesus gave to those who complained about his failure to observe the sabbath – an observance, Augustine points out, that is based on the authority of the Genesis passage on which he is commenting. But he qualifies that the observance of the sabbath was a "shadow of what was to come [Heb 10:1], to prefigure the spiritual rest which God was promising in a kind of secret code, on this model of his own rest, to the faithful doing their own good works."[46] Further, the mystery of this rest was shown in the burial of Christ. Jesus was in his tomb on the sabbath, spending the entire day as a kind of "holy vacation, after he had finished all his works on the sixth day, that is Preparation Day,[47] which they call the sixth of the Sabbath, when what had been written about him was fulfilled on the very gibbet of the cross."[48] In a move that ties together the days of Genesis, sabbath, and rest, Augustine sees the last words of Christ on the cross ("It is finished") as connected to the day of rest inaugurated by God in the text of Genesis. "So why," he insists, "should we be surprised if God wished to point forward to this day on which Christ would rest in the grave, and to this end rested from his works on one day, before

---

[44]Ibid.

[45]As cited by Augustine in ibid., 4.11.21 (TWSA I/13:252).

[46]Ibid., 4.11.21 (TWSA I/13:253).

[47]In a footnote the translator, Edmund Hill, states that this is likely the phrase used in Christian circles of Augustine's time to indicate what we call Good Friday. See TWSA I/13:253n23.

[48]Ibid., 4.11.21 (TWSA I/13:253).

proceeding from then on to work the unfolding of the ages, in order to verify these other words too: *My Father is working until now?*"[49]

Augustine also offers another way of understanding God's resting and continuous working – resting from establishing any new kinds of creatures. In this understanding God's working is his providential caring of what he had set in place, for surely he did not withhold on that seventh day his "power . . . from the government of heaven and earth and of all things he established."[50] In fact, had this indeed happened, it all would have collapsed into nothingness. God's creation is not like a house, which continues to stand after the builder finishes his work and walks away. If God withdraws his providential caring from his creation it would "not be able to go on standing for a single moment."[51] Thus, the expression "My Father is working until now" demonstrates this continuous providence.

We should recall here how Augustine earlier connected the seventh-day rest in Genesis with Jesus' answer to those who were complaining that he failed to observe the sabbath in John 5:17.[52] He concludes that he has satisfactorily shown that both scriptural statements (that God rested on the seventh day and that he is working until now) are true. He now reminds us of that same passage in John and its connection to the sabbath of Genesis. God commanded the Jewish people to observe the sabbath in order to "represent and point to" the rest that we shall see after our own good works. Unfortunately, they obeyed the command in a "literal-minded, materialistic way," which was why they protested Jesus' failure to observe the sabbath.[53] But now, "in the time after the full revelation of grace," the observance of that sabbath is eliminated because in this present time of grace we work in the hope of a future sabbath that will be perpetual.[54]

---

[49]Ibid. Italics in original translation.
[50]Ibid., 4.12.22 (TWSA I/13:253).
[51]Ibid.
[52]See ibid., 4.11.21 (TWSA I/13:252-53).
[53]Ibid., 4.13.24 (TWSA I/13:255).
[54]Ibid.

Since God himself neither grew tired when he created nor found relief when he stopped, why did he sanctify this seventh day? According to Augustine, God urged us to rest by "intimating to us that he had sanctified the day on which he rested from all his works."[55] God did not sanctify any of the previous six days, not even the sixth on which he created humans. But he sanctified the seventh – why? In his answer, Augustine proceeds on the foundation of "two indubitable truths":

> that God did not delight in some kind of temporal period of rest after hard toil and the much desired end of his business; and that it was not falsely or vainly that these scriptures, so deservedly surpassing all others in their authority, said that God rested on the seventh day from all the works which he had made, and for that reason sanctified this same day.[56]

God himself is presented as a model by his resting from all the works that he had made. According to Augustine, God did not delight in any of the things he created in a way that suggested he really needed them. No, God delighted in himself, just as we should delight in him and not our good works. "God's rest therefore, to those who understand it correctly, means his being in need of no one else's good; and for this reason his resting in us is certainly in himself, because we too find bliss in the good which he is, not he in the good which we are."[57]

God's rest, then, comes after the "six days" of work in the Genesis account. The seventh day has a beginning (morning) but no end (evening). In the days of creation indicated in Genesis, evening represented the completion of creation for that day, and morning indicated the starting point for the creation to be fashioned on the next day. The seventh day, however, has morning but no evening. Augustine explains, "Because nothing else remained to be fashioned, there was made morning after that [sixth] evening in such a way that it would not be the starting post for fashioning another creature, but the start of quiet

---

[55]Ibid., 4.14.25 (TWSA I/13:255).
[56]Ibid. (TWSA I/13:256).
[57]Ibid., 4.16.27 (TWSA I/13:257).

rest for the universal creation in the quiet rest of the creator."[58] The morning of the seventh day did not indicate the starting point for creation to be fashioned on that day, as it did the other six. Rather, it indicated the beginning of its abiding in the rest of God. He continues, "And that is why the seventh day begins for this creation with a morning, but does not terminate in any evening."[59]

Every other day had a morning and evening, but Augustine continues to ponder why the seventh did not have an evening. "After all it too is one of the days which make up the seven, by whose regular repetition the months and years and ages run their course."[60] But if this had happened, Augustine explains, there would have been an eighth day. There was no indication of the close of the seventh day because the eighth day "is the same as the first" – that is, "this weekly series continues to roll." So the seven-day period continually recurs in time, but the first six days "were unfolded in a manner quite beyond what we are used to in our experience."[61] Augustine makes this claim because of some of the strange things he notices in the Genesis account that are not in keeping with what we observe about days. For example, the first three days of creation in Genesis occurred "before the fashioning of the lamps in the sky."[62] But, admitting that he is quite unclear about what kind of evening and morning were on those first six days, he is certain that God's rest did not start after the evening of the sixth day. If that were the case, then time would affect God, leading to the assumption that something starts and finishes for him. The quiet rest of God does not have a starting point. He finds "rest in himself and bliss in the good for which he is for himself, [it] neither starts for him nor terminates."[63]

---

[58]Ibid., 4.32 (TWSA I/13:259).
[59]Ibid.
[60]Ibid., 4.33 (TWSA I/13:259).
[61]Ibid.
[62]Ibid.
[63]Ibid., 4.34 (TWSA I/13:260).

In a move that could be described as sacramental, Augustine then discusses the goal of the "whole universe of creation" in relation to God's rest. The universe of creation that was completed in six days has an end in its own nature. But it also serves another goal, which it has in the rest of God. The proper end and goal of all creation is not found in itself but in the one who created it:

> And for this reason, while God abides in himself, he swings everything whatever that comes from him back to himself, like a boomerang, so that every creature might find in him the final terminus and goal for its nature, not to be what he is, but to find in him the place of rest in which to preserve what by nature it is in itself.[64]

For Augustine, the morning that was made after the evening of the sixth day represents "the start of creation's sharing in the quiet rest of the creator; after all it could not rest in him unless it had been perfected."[65] The lack of evening on the seventh day, therefore, represents the whole created universe always abiding in its Creator.

Augustine still wonders what this seventh day is – a creation by God or a period of time? As we learned earlier, he argues that time is created. Therefore, if the seventh day is a time, he recognizes that it must have been created by God. The text of Genesis indicates what creatures were created on the other six days, so we know when they were created. But it is not so with the seventh day. Augustine asks, "How did God rest on a day which he had not created?"[66] God created only one day, which repeated itself, thus leaving no need to create a seventh day because it was already made by the simple repetition of this one day. In this way of thinking the seventh day on which God rested was the seventh repetition of the day that was originally created. Thus, the seventh day is not a creature on its own "but the same one coming round seven times, the one which was fashioned when God called the light day and the darkness night."[67]

---

[64]Ibid.
[65]Ibid., 4.35 (TWSA I/13:260).
[66]Ibid., 4.20.37 (TWSA I/13:262).
[67]Ibid.

THE CREATION OF LIGHT IN
THE LITERAL MEANING OF GENESIS

This leads Augustine to consider the light that was created on day one. In book 1 of *The Literal Meaning of Genesis* he discussed how light could be involved in the alternations of day and night before the sun was created.[68] In book 4 he reminds readers of his conclusion about the light of the first three days:

> We should say the light created originally is the forming and shaping of the spiritual creation, while the night is the material of things still to be formed and shaped in the remaining works, material that had been laid down when in the beginning God made heaven and earth, before he made the day by a word.[69]

Thus, the light that was originally created is a "spiritual, not a bodily light."[70] How is this spiritual light related to the evenings and mornings that are indicated in the Genesis account? Augustine has already told us that the creation of light in Genesis 1:3 refers to the illumination of the angels. He reasons that the claim in Genesis 1:1 that God made the heavens must include the angels, but since no lapse of time is involved, they must be in an unformed state until God illumines them.[71] The spiritual light (the angels), Augustine explains, was created after the darkness that was over the abyss. He understands this as meaning its transition from its unformed state toward its formation by the Creator. In a similar manner, morning was made after evening, which means that once it knows its own nature (that it is not God), it returns to praise God, who is the true light who formed it in the first place. The creatures below the angels, which are the created spiritual light, are all made with the knowledge of the angels; "that surely is why the same day is repeated every time, so that by its repetition as many

---

[68]This issue has been briefly discussed above in relation to *City of God* 11.7. Augustine addresses it in more detail in *Literal Meaning of Genesis* 1.4.9–5.11; 1.10.19-22; 1.14.28-30; 1.17.32-35.

[69]Augustine, *Literal Meaning of Genesis* 4.21.38 (TWSA I/13:263).

[70]Ibid., 4.22.39 (TWSA I/13:263).

[71]Ibid., 1.17.32 (TWSA I/13:183-85).

days may be made as there are distinct kinds of things created, to be brought to an end with the perfection of the number six."[72]

This means that the evening of the first day is knowledge that spiritual beings have of themselves, that they are creatures. Morning marks the end of one day and the start of the second day. In this spiritual understanding, morning means the spiritual creation turning to praise the Creator and receiving from the Word of God knowledge of what it will be made next – namely, the "solid structure," or firmament. Thus, creation occurs first in knowledge of the spiritual beings and then in the actual firmament itself. And so it continues in this manner for each "day."

> After this there is made morning, which concludes the second day and starts off the third; this morning again marks the conversion of this light in the same way, that is, the turning of this day to the praise of God for his work of making the solid structure, and to its reception from his Word of knowledge of the creature to be fashioned after the solid structure.[73]

If we understand this quotation in the context of what was just stated about creation being in the knowledge of the angels before it was manifested, we can better understand what Augustine is saying here, and the temporal importance of the days fades in the distance. The knowledge of the angels being connected to evening and morning simply repeats itself in the days of creation. Morning indicates knowledge of their own spiritual "higher" order, albeit not what God is, while evening indicates a "lesser degree of knowledge" – that is, a knowledge of the lower order of creation. For Augustine, knowledge of a thing in the Word of God is "day," while knowledge of its own specific nature is "evening."[74]

Angels, who always see the face of God and enjoy the Word of God, were the first to come to know the universal creation because they were the first created in the Word of God himself. In the Word of God there

---

[72]Ibid., 4.23.39 (TWSA I/13:264). On the perfection of the number six, see *Literal Meaning of Genesis* 4.2 (TWSA I/13:242-45).

[73]Ibid.

[74]Ibid., 4.23.40 (TWSA I/13:264).

exists the "eternal ideas [*rationes*]"of the things that were made because through him all things were made.[75] The angels know these eternal ideas before they are created in time. When they are created, they turn back "to the praise of the one in whose unchangeable Truth they originally see the ideas according to which [the universe] was made."[76] The knowledge the angels have of the spiritual, eternal ideas is like the day, while the knowledge they have of time-bound creation is like the evening. But since angelic knowledge does not dwell on what has been created in time without praising God for it, morning comes (which is repeated on all six days) with a knowledge of what will be created.

So, the whole created universe was completed and finished by six repetitions of the originally created day. The morning from which the seventh day began did not have an evening "because rest is not part of God's creation."[77] When creatures were being made on the other days, the angels knew them in their eternal ideas and in their actuality in time. It is this creation in time that is evening. This means that the day is not to be understood as the "form of the actual work"; the evening is not to be understood as the day's end nor the morning its beginning.[78] If we concluded that these things were true of the day, "we would be compelled to say against scripture either that the seventh day was fashioned as a creature outside the six days, or else that the seventh day was not itself something created."[79] Instead, the day is repeated through God's works – not in "bodily circular motions" but in "spiritual knowledge," where the angels contemplate in the Word of God that about which God says "Let it be made." It is in the angelic knowledge that things are first made, and after that "do they know the actual thing made in itself, which is signified by the making of evening. They then refer this knowledge of the thing made to the praise of that Truth

---

[75]Ibid., 4.24.41 (TWSA I/13:265).
[76]Ibid.
[77]Ibid., 4.26.43 (TWSA I/13:266).
[78]Ibid.
[79]Ibid., 4.26.43 (TWSA I/13:267).

where they had seen the idea of making it, and this is signified by the making of morning."[80]

Augustine claims that this way of understanding things allows for those three days that were mentioned before the creation of the sun and the moon to give light. There is, he states, just one day, and it should not be understood in the same way we understand days that are measured and counted by the sun's circuit. The day that was originally repeated three times before the creation of the sun and moon on the fourth repetition is not the same kind of day we experience. Thus, the night and day that God divided in Genesis 1:4 "are to be taken in quite a different sense from this night and day, between which he said that the lamps he created were to divide, when he said, *And let them divide between day and night* (Gen 1:14)."[81] The fourth day was when God fashioned the kind of day we know. But the day that was originally created had already gone through three repetitions before the lights were created on the fourth repetition.

We have no perception of or experience with the kind of days that Augustine is setting forth from the text of Genesis. Because of this he advises that we not "rush into any assertion of any ill-considered theory about them."[82] He emphasizes that these seven days of creation are not at all like "the week that whirls times and seasons along by its constant recurrence."[83] Augustine insists that his understanding of the days is not "figurative and allegorical" or "metaphorical." "Certainly," he claims, "it is different from our usual way of talking about this bodily light of every day, but that does not mean that here we have the strict and proper, there just a metaphorical, use of these terms."[84]

In the days we experience, light declines as we get closer to sunset, and we call that evening. Also, light returns as we get closer to sunrise,

---

[80]Ibid.
[81]Ibid. Italics in original translation.
[82]Ibid., 4.27.44 (TWSA I/13:267).
[83]Ibid.
[84]Ibid., 4.28.45 (TWSA I/13:268).

and we call that morning. But since we have a "surer light," we also have a "surer day" and, therefore, "both a truer evening and a truer morning."[85] It thus makes perfect sense to Augustine that a spiritual evening occurs when there is a turning away from contemplating the Creator, and a spiritual morning when there is a move from knowledge of the Creator to praise of him. For Augustine, this is actually a *literal* interpretation, not allegorical. He recognizes that some may not be satisfied with "the line which I have been able in my small measure to explore or trace."[86] He encourages those who disagree to find another explanation, but it must be "as a strict and proper account of the way the foundations of this creation were made."[87] In other words, it must also be a *literal* interpretation.[88]

SIMULTANEOUS CREATION

The days Augustine speaks of were not separated by intervals of time; rather, it was "a matter of the spiritual power of the angelic mind, comprehending all that it wished in a simultaneous knowledge with the utmost facility."[89] The human mind, even if it is "fervent and eager" to know, simply cannot comprehend in this simultaneous manner. But the angelic mind was created before all and is "adhered in unalloyed charity to the Word of God."[90] The angelic mind saw the things to be made in the Word of God before they were made – "before being made in their own proper nature, they were first made in the knowledge of that mind, when God said, 'Let them be made.'"[91]

Even though the text of Genesis puts things one after another in a chain of causes, the angelic mind is able to grasp things simultaneously.

---

[85]Ibid.
[86]Ibid.
[87]Ibid.
[88]St. Augustine, *The Literal Meaning of Genesis*, vol. 1, *Books 1–6*, trans. John Hammond Taylor, Ancient Christian Writers 41 (New York: Paulist Press, 1982), 136n59.
[89]Augustine, *Literal Meaning of Genesis* 4.32.49 (TWSA I/13:270).
[90]Ibid.
[91]Ibid.

But, Augustine wonders, does this mean that all creation was made simultaneously rather than at intervals according to the predetermined days? He argues that the act of creation was not governed by the same natural motions we now experience and observe; rather, it was "fashioned according to the wonderful and inexpressible virtuosity of the Wisdom of God, *who reaches from end to end mightily, and disposes all things sweetly* (Wisdom of Solomon 8:1). She, after all, does not reach there by steps, or arrive as if by pacing out the distance."[92] Augustine believes that creation occurred "from those implanted formulae or ideas [*rationes*] which God so to say scattered like seed in the very moment of fashioning them."[93]

These "ideas" (*rationes*) should be distinguished from the abovementioned ideas that were said to be contemplated by the angels in the Word of God and seen as divine exemplars of the works that are created. Here Augustine means causal reasons implanted by God in the created world that account for the growth of beings that appear throughout the ages. This has come to be called Augustine's seminal principles. These are "seed-like powers in the created world, causing the seeds to develop according to God's plan."[94] They are the created manifestations of the uncreated ideas that are "programmed into" creation.[95]

Creation for Augustine did not occur in a "time-measured way."[96] Plants require time to root and germinate under the soil before they break forth from the earth, so they would have needed much more than a day to mature. How many days would it take for a bird to start flying if it grew in a natural way to develop the feathers and wings that are proper to its nature? Maybe it was only the eggs that were created when, on the fifth day, the waters brought forth birds. Perhaps everything grew in a definite number of days because there were already

---

[92]Ibid., 4.33.51 (TWSA I/13:272). Italics in original translation.
[93]Ibid.
[94]Taylor, *Literal Meaning of Genesis*, 141n67.
[95]TWSA I/13:173n47.
[96]Augustine, *Literal Meaning of Genesis* 4.33.52 (TWSA I/13:272).

implanted these seeds, "being woven incorporeally into the texture of corporeal things."[97] In the liquid element itself was this "formulae [*rationes*] . . . according to which the fowls of the air would be able to arise and reach full growth and perfection through the periods of time proper to each kind."[98]

We are told in Genesis how God finished his work in six days, but Scripture elsewhere tells us that he "created all things simultaneously together."[99] For Augustine, we should not see this as a contradiction because this one day repeated six or seven times was made simultaneously. The reason why Genesis so "distinctly and methodically" recounts the days is "for the sake of those who cannot arrive at an understanding of the text, 'he created all things together simultaneously,' unless scripture accompanies them more slowly, step by step, to the goal to which it is leading them."[100]

Even though both passages of Scripture above are true, Augustine believes it is actually easier to understand simultaneous creation than to speak about it as a sequence. To clarify, he employs an example of our observation of the sunrise. Our gaze reaches the sun by passing through a significant expanse of air and sky. Our glance crosses air above land and bodies of water like lakes and rivers. We speak about our gaze crossing land *before* it crosses water and also crossing more land *after* it crosses the water. Just because the terms *before* and *after* were used in this kind of description, does that mean "our glance does not cross all these spaces instantaneously, that is simultaneously, in

---

[97]Ibid., 4.33.52 (TWSA I/13:273).

[98]Ibid.

[99]Ibid. The passage referred to here is from Wisdom of Sirach, or Ecclesiasticus, 18:1. Augustine, along with most other church fathers, considered this book as Scripture. See Craig Allert, *A High View of Scripture? The Authority of the Bible and the Formation of the New Testament Canon*, Evangelical *Ressourcement* Series (Grand Rapids: Baker Academic, 2007). In *Literal Meaning of Genesis* 5.1.1–5.3.6 Augustine argues that Gen 2:4 ("These are the generations of the heavens and the earth when they were created") clearly indicates simultaneous creation and "signals the inception of the actual, empirical creation." See Paul Blowers, *Drama of the Divine Economy: Creator and Creation in Early Christian Theology and Piety*, Oxford Early Christian Studies (Oxford: Oxford University Press, 2012), 155.

[100]Ibid.

one single blink?"[101] When someone turns toward the sun with her eyes shut, she will expect to see the sun as soon as she opens them. In fact, it will seem that she sees the sun even before her eyes are fully opened. This is a simultaneous occurrence.

Augustine argues that this is why Paul, in 1 Corinthians 15:52, expressed the speed of our resurrection as happening "in the blink of an eye." In the material realm, nothing more rapid can be found. If the glance of an eye is capable of such speed, surely the human mind must be capable of more, and the angelic of even more. But ultimately,

> what now must be said of the swiftness of that supreme Wisdom of God, who *reaches everywhere on account of her purity, and nothing defiled affects her* (Wisdom of Solomon 7:24-25)? So then, in these things that were made all together simultaneously, nobody can see what should be made before or after which, except in that Wisdom through whom they were made simultaneously, and in order.[102]

In his conclusion to book 4 of *The Literal Meaning of Genesis*, Augustine summarizes simultaneous creation:

> So then, this day which God made in the first place – if it is the spiritual and rational creation, that is, the supercelestial angels and powers – was presented with all the works of God in a sequence or order which is the order of knowledge. Thus it would foreknow in the Word of God the things that were to be made, and then know them as creatures already made; this, however, would not be at distinct moments of time, but would be a knowledge of before and after in the sequence of creatures, of all at once simultaneously in the effective work of the creator. For he did indeed make things that were going to be in the future; but he made time-bound things in a timeless manner, to let times made by him run their proper course.[103]

The seven days that we experience are a "shadowy resemblance" to the days described above. Our days are meant to remind us to seek "those

---

[101]Ibid., 4.34.53 (TWSA I/13:274).
[102]Ibid., 4.34.55 (TWSA I/13:275). Italics in original translation.
[103]Ibid., 4.35.56 (TWSA I/13:275).

days in which the created spiritual light was presented with all the works of God through the perfection of the number six, and was then able to have a morning, but not an evening, on the seventh day of God's rest."[104]

For Augustine there are two phases of creation – the primary simultaneous creation of all things and the administration of creation in space and time. These two phases culminate in God's rest and the Father's providence – themes we have witnessed in the preceding pages. During his discussion of God's rest in book 4 of *The Literal Meaning of Genesis*, he reflects on how rest can apply to God, who really cannot be said to require rest. In fact, he asks his readers to consider the inauguration of God's rest in Genesis 2:1-3 in light of John 5:17: "My Father is working until now, and I myself am working."[105] This contrast between God's transcendent rest on the seventh day and Jesus' own testimony that his Father is still working is Augustine's most forcible scriptural justification for these two phases of creation. Thus the completion of creation is marked by the divine rest, and the providential plan of the Creator in caring for and preserving the world is marked by Jesus' words in the Gospel of John.[106]

LIGHT IN *THE CITY OF GOD*

This brings us back to the passages of our original focus, *The City of God* 11.8 and 22.30. We should recall that in 11.7 Augustine points out how difficult it is to understand the days of Genesis. He wonders how, if the sun was created on the fourth day, three days can pass by. If the light mentioned in Genesis 1:3-4 was not the sun, then what was it? He then begins his argument for the light being the holy city of spiritual beings. He begins by comparing the six days of creation with the creature's knowledge. Creaturely knowledge is like the "dusk of the evening," but when it is turned in praise and love to the Creator "it becomes the full

---

[104]Ibid.
[105]As cited by Augustine in *Literal Meaning of Genesis* 4.11.21 (TWSA I/13:252).
[106]Blowers, *Drama of the Divine*, 155.

light of the morning."[107] He applies this as a sort of template to each of the days of creation:

> And when it does this in its knowledge of itself, this is the first day. When it does this in its knowledge of the firmament, which separates the lower waters from the upper waters and is called the heaven, that is the second day. When it does this in its knowledge of the earth and the sea and all that grows from the earth and has its roots in it, that is the third day. When it does this in its knowledge of the greater and lesser lights of heaven and all the stars, that is the fourth day. When it does this in its knowledge of all the living things that swim in the waters and of all that fly, that is the fifth day. And when it does this in its knowledge of all the beasts of the earth and of man himself, that is the sixth day.[108]

On the seventh day God rested, which, here in *The City of God*, Augustine equates with us resting in God.

We soon learn, however, that the creaturely knowledge to which Augustine has been referring is actually angelic knowledge. The angels, Augustine argues, form the greater part of the holy city to which he referred in the previous chapter. Scripture does not tell us about the creation of angels. As we saw above, Augustine believes that the creation of angels is signified by the creation of light in Genesis 1:3.[109] Even though the creation of angels is not expressly included in the works of God from which he rested on the seventh day, Augustine insists that they must have been created. After all, other parts of Scripture indicate that they are creatures.[110]

Augustine also sees support in Scripture for when the angels were created: "Who would dare to suppose that the angels were not made until after all the things enumerated on the six days of creation?"[111] This kind of "foolish nonsense" is refuted by Job 38:7: "When the stars were

---

[107]Augustine, *City of God* 11.7 (TWSA I/7:8).

[108]Ibid.

[109]Ibid., 11.9 (TWSA I/7:9).

[110]Augustine cites Dan 3:57 (LXX), where angels are listed as the "works" of God, and Ps 148:1-5 for support.

[111]Augustine, *City of God* 11.9 (TWSA I/7:10).

made, all my angels praised me with a loud voice."[112] From this he deduces that angels were created before the stars, which were created on day four. But this does not mean the angels were created on the third day, because we already know from Scripture what was created on that day. The same goes for day two. Augustine continues, "Clearly, then, if the angels are included in the works of God on those six days, they are the light which was given the name *day*, and to stress the unity of that day, it was called not 'first day' but rather *one day*."[113] The angels are participants in the eternal light, which Augustine connects with the wisdom of God, through which all things were made. This, of course, is none other than "the only begotten Son of God."

> Thus the angels, illumined by the light that created them, themselves became light and were called *day* by virtue of their participation in the immutable light and day which is the Word of God, by which both they and everything else were created. For *the true light that enlightens everyone coming into this world* (Jn 1:9) also enlightens every pure angel, so that he is light not in himself but in God.[114]

We should therefore understand Augustine's explanation of creaturely knowledge in *The City of God* 11.7 as angelic knowledge. In fact, what he says about angelic knowledge in this passage is very similar to what he says about it in *The Literal Meaning of Genesis*.[115] Before things are created in time, angels arrive at knowledge of the eternal ideas of things made through the Word of God. Augustine says that this knowledge is like the day. The actual creation of the eternal ideas causes the angels to turn back "to the praise of the one in whose unchangeable Truth they originally see the ideas according to which it was made."[116] The knowledge the angels have leads them to praise God, rather than that which is created by God in time, which is then the

---

[112]As cited by Augustine.
[113]Ibid., 11.9 (TWSA I/7:10). Italics in original translation.
[114]Ibid. Italics in original translation.
[115]Augustine, *Literal Meaning of Genesis* 1.17.32; 4.23.39-41.
[116]Ibid., 4.24.41 (TWSA I/13:265).

return of morning. This is repeated on all six days. Thus, as Augustine concludes in *The City of God*:

> Clearly, then, if the angels are included among the works of God on those six days, they are the light which was given the name *day*, and, to stress the unity of that day, it was called not "the first day" but rather *one day*. Nor is the second a different day, or the third, or the rest. Rather the very same one day is repeated up to the number six or seven on account of the seven phases of knowledge, namely, the six of the works that God created and the seventh of God's rest.[117]

THE EIGHTH DAY

This brings us back to the topic of God's rest in the final chapter of *The City of God*, book 22.30. Daniélou calls this passage "the most beautiful that has been devoted to the spiritual Sabbath."[118] The context has Augustine writing about the heavenly city in which believers will receive their eternal rest, where the words of Psalm 46:10 will be fulfilled.[119] He calls this the "supreme Sabbath," which does not end. Drawing on the observation that Moses does not mention evening on the seventh day, he cites Genesis 2:2-3[120] and explains that "we ourselves shall be the seventh day, when we have been made full by his blessing and made new by his sanctification. . . . Then we shall be still and see that he is God. . . . But when we are made new by him and are perfected by his greater grace, we shall be still for all eternity and see that he is God, and by him we shall be filled when he is all in all."[121]

Augustine claims that we will be able to understand the sabbath about which he writes "if we enumerate the ages, as if they were days,

---

[117]Augustine, *City of God* 11.9 (TWSA I/7:10). Italics in original translation.

[118]Jean Daniélou, *The Bible and the Liturgy*, Liturgical Studies 3 (Notre Dame, IN: University of Notre Dame Press, 1961), 283.

[119]Ps 46:10: "Be still and see that I am God."

[120]See *City of God* 11.6; *Confessions* 13.36.51; *The Literal Meaning of Genesis* 4.18.31; and *On Genesis* 1.41. Gen 2:2-3: "And on the seventh day God rested from all his works that he had done. And God blessed the seventh day and sanctified it, because on it he rested from all his works that he had begun to do."

[121]Augustine, *City of God* 22.30 (TWSA I/7:553).

according to the divisions of time that we see represented in Scripture, for we shall find that it is the seventh age."[122] This connection between the seven days of Genesis and a division of seven ages is made elsewhere in Augustine's writings, and it is usually in parallel reference to the concept of sabbath resting.[123] This enumeration of the ages corresponding to the days of creation and culminating in the day/age of rest is well established in his own thinking. In this passage of *The City of God*, he describes the ages this way:

> The first age, counted as the first day, extends from Adam to the flood, and the second from the flood to Abraham. These two are equal not in length of time but in number of generations, for each turns out to contain ten generations. From Abraham down to the coming of Christ, by the reckoning of the evangelist Matthew, there are three ages, each extending across fourteen generations – one from Abraham to David, the second from David to the exile to Babylon, and the third from the exile to Babylon down to Christ's birth in the flesh. These make five ages in all. The sixth age is now in progress, it is not to be measured by any set number of generations, for Scripture says *It is not for you to know the times that the Father has put in his own power* (Acts 1:7). After this sixth age God will rest, as on the seventh day; and he will cause this same seventh day – the day that we ourselves shall be – to rest in him. It would take too long, however, to discuss each of these ages in detail now. It is enough simply to point out that this seventh age will be our sabbath, and its end will not be an evening but rather the Lord's Day, as an eighth and eternal day, consecrated by Christ's resurrection, and prefiguring the eternal rest not only of the spirit but also of the body. There we shall be still and see, see and love, love and praise. Behold what will be in the end without end! For what else is our end but to reach the kingdom that has no end?[124]

Here Augustine makes reference to "an eighth and eternal day," with which we are familiar from Basil. We should recall the early church's

---

[122]Ibid., 22.30 (TWSA I/7:554).

[123]See, for example, *Answer to Faustus, a Manichean* 12.8; *Confessions* 13.3651–37.52; *Instructing Beginners in Faith* 17.27-28; *Miscellany of Eighty-Three Questions* 58.2; *On Genesis* 1.23.35–24.42; Homily 259; and *The Trinity* 4.4.7.

[124]Augustine, *City of God* 22.30 (TWSA I/7:554). Italics in original translation.

celebration of the Easter Octave, which was the first Sunday after Easter. The name *Octave* means eight because the celebration was eight days after Easter. This octave structure (first day → six days → eighth day) became very important in the liturgy and the spiritual formation of Christians.[125] To put Augustine's comments on the seventh day/age and the eighth day in context, it is helpful to look at Augustine's Homily 259, "For the Octave of Easter."[126]

Augustine's Homily 259 (ca. 394) is preached in this context of the Easter Octave. His very first statement reveals both the eighth day's significance and its eternality: "This day is a symbol of perpetual joy for us, for the life which this day signifies will not pass away as this day is going to pass away."[127] As we just saw above in *The City of God*, Augustine makes a strong connection between the seventh age, which is the sabbath that has no end ("no evening") and is called the Lord's Day, and an eighth day that is eternal. This prefigures the eternal rest of the spirit and body.[128] The eternality of the eighth day is emphasized in the homily as well. This is the life Christians will enjoy with the angels – there will be perpetual peace and happiness that lasts forever, and there will be no anxiety, sadness, or death. Then Augustine explicitly connects this as the eighth day: "Therefore, the eighth day signifies new life at the end of the world."[129] Further, in order to explain the eighth day, he appeals to his previously mentioned division of ages that are symbolized by the days of creation:

> That will be the seventh day, just as if the first day in the whole era were the time from Adam to Noe; the second, from Noe to Abraham; the third, from Abraham to David, as the Gospel of Matthew divides it [Mt 1:17]; the fourth, from David to the Transmigration into Babylon; the fifth, from the Transmigration to the coming of our Lord Jesus Christ. The sixth day,

---

[125]Blowers, *Drama of the Divine Economy*, 343.
[126]All quotations from Homily 259 are from Saint Augustine, *Sermons on the Liturgical Seasons*, trans. Mary Sarah Muldowney, FC 38 (New York: Fathers of the Church, 1959), 368-77.
[127]Ibid., 259.1 (FC 38:368).
[128]Augustine, *City of God* 22.30 (TWSA I/7:554).
[129]Augustine, Homily 259.2 (FC 38:368).

therefore, begins with the coming of the Lord, and we are living in that sixth day. Hence, just as in Genesis [1:27] [we read that] man was fashioned in the image of God on the sixth day, so in our time, as if on the sixth day of the entire era, we are born again in baptism so that we may receive the image of our Creator. But, when that sixth day will have passed, rest will come after the judgment, and the holy and just ones of God will celebrate their sabbath. After the seventh day, however, when the glory of the harvest, the brightness and the merit of the saints have appeared on the threshing floor, then we shall go to that life and rest of which the Scripture says: "Eye has not seen nor ear heard, nor has it entered into the heart of man, what things God has prepared for those who love him [1 Cor 2:9]." Then we return, as if to the beginning, for just as when seven days have passed, the eighth becomes the first [of a new week], so after the seven periods of this transitory world have been spent and completed, we shall return to that immortal blessedness from which man fell. Hence, octaves complete the sacraments of the newly baptized.[130]

In Homily 260c Augustine compares the promises of Isaiah 57:19, "peace upon peace," with the sabbath day of rest that is contained "in this temporal round of days."[131] God rested on the seventh day in order to indicate the eternal rest of his saints. This is foretold in Job 5:19, "He will deliver you from six troubles; in seven no harm shall touch you." The reason Genesis does not indicate an evening on the seventh day is because it moves into the eighth day of eternity. But the eighth day is not the only thing that should be an indicator of eternity for Augustine. He chides "lovers of this world" who do not consider the "symbolic meaning of the days." Failure to do so shows that their focus is "not the rest of a spiritual sabbath, from which their thoughts could also be directed to

---

[130]Augustine, Homily 259.2 (FC 38:370). Daniélou (*Bible and the Liturgy*, 277) notes that this passage shows a definite millenarianism. That is, there is the idea of a glorious reign of Christ and his saints here on the earth, which corresponds to the seventh age. This reign is before the eighth day, which is everlasting life. While this is an interesting issue in itself, it should not divert us from the reason this passage interests us – because of the octave structure and its connection to the days of Genesis.

[131]Augustine, Homily 260c.4 (TWSA III/7). All quotations from Homily 260c are from Saint Augustine, *Sermons on Liturgical Seasons*, trans. Edmund Hill, TWSA III/7 (Hyde Park, NY: New City Press, 1993).

the eternity of the eighth." Rather, they are "given over . . . to the round of temporal thoughts, unable to entertain any idea of the eternal."

"IN THE BEGINNING"

If we comprehend the important distinctions Augustine has made between time and eternity, we are in a better position to understand the irony when we say that a created world "always" existed or that there was no time before the world. Time was created with the world, so to speak of time outside of this context is meaningless. God is prior to the creation of the world by his eternity, not by precedence. And, in this sense, eternity does not mean an indefinite extension of time.[132] It now remains for us to return to the initial investigation of this chapter – to see how Augustine understood the phrase "In the beginning."

In book 1 of his *Literal Meaning of Genesis*, Augustine refers to the twofold structure of divine Scripture – the two testaments. In the context of one of the best-known comments by Augustine on the book of Genesis, he calls the reader to seriously consider "what eternal realities are there suggested, what deeds are recounted, what future events are foretold, what actions commanded or advised."[133] This is an exhortation to read and interpret the Scriptures as divine. When one considers accounts of things done in the Scriptures, Augustine asks whether they should be taken as only having a "figurative" meaning or whether they should also be seen as a "faithful account of what actually happened."[134] Then he offers his famous comment:

> No Christian, I mean, will have the nerve to say that they should not be taken in the figurative sense, if he pays attention to what the apostle says: *All these things, however, happened among them in figure* (1 Cor 10:11), and to his commending what is written in Genesis, *And they shall be two in one flesh* (Gn 2:24), *as a great sacrament in Christ and in the Church* (Eph 5:32).[135]

---

[132]Christian, "Augustine and the Creation of the World," 7-8.
[133]Augustine, *Literal Meaning of Genesis* 1.1.1 (TWSA I/13:168).
[134]Ibid.
[135]Ibid. Italics in original translation.

This is certainly an odd way to start off a commentary on the *literal* meaning of Genesis! While acknowledging the physical, twofold structure of Scripture into testaments, Augustine apparently sees a more significant division between a figurative approach and an approach that reads it as an account of "what actually happened." Further, while acknowledging that the text "has to be read in both ways," it appears that Augustine has a preference for the figurative reading because it opens the mysteries of Christ to the reader. He even uses Paul ("the apostle") as his authority and predecessor in reading Scripture this way.[136]

The hint of Augustine's preference for a figurative reading is confirmed when he begins to address what the word *beginning* means in Genesis 1:1. He insists that the text must be treated in both ways and asks what the phrase "In the beginning God made heaven and earth" means "apart from its allegorical significance."[137] He offers three choices: the beginning of time; the first of all things; or the Word of God, the only begotten Son. His conclusion hinges, once again, on the eternality of God – the concept we unpacked from Augustine in the previous pages of this chapter.

Augustine connects questions about the meaning of *beginning* in Genesis 1:1 with "Let light be made" in Genesis 1:3. He asks the same question about "Let light be made" as he does about "In the beginning God made heaven and earth." Was the statement made in time or in the eternity of the Word?[138] He concludes that if the command "Let light be made" was in time, then it involved change. Therefore, the statement was not made in time, because God is "not subject to change."

Augustine is using the example of God's command "Let there be light" to inquire about how God "says" things through his Word. If the voice was an audible command, what language did God use?[139] There was no

---

[136]Ibid., 1.1.2 (TWSA I/13:168).
[137]Ibid.
[138]Ibid., 1.2.4 (TWSA I/13:169).
[139]Ibid., 1.2.5 (TWSA I/13:170).

diversity of languages yet because, according to him, that occurred only at the building of the tower after the flood (Gen 11:1-9). Further, who was there to hear and understand the audible utterance? Augustine concludes that this line of thought is not very profitable, asking rhetorically, "Is this an altogether absurd and literal-minded, fleshy, train of thought and conjecture?"[140] His answer, apparently, is yes.

So then, what is the voice of God? How does God "say" things? This brings Augustine to explicitly connect Genesis 1:1 with John 1. Perhaps the voice of God is not an audible utterance but belongs "to the very nature of his Word, about which it is said, *In the beginning was the Word, and the Word was with God, and it is God that the Word was* (Jn 1:1)."[141] All things were made by the Word (Jn 1:3), so it is clear that light was also made through this eternal Word.

> If that is the case, then God's saying *Let light be made* is something eternal, because the Word of God, God with God, the only Son of God, is co-eternal with the Father, although when God said this in the eternal Word, a time-bound creature was made. While "when" and "some time" are time words, all the same the time when something should be made is eternal for the Word of God, and it is then made when it is in that Word that it should have been made, in the Word in which there is no "when" or "some time," because that whole Word is eternal.[142]

Augustine then turns his attention to inquiring what this light was — something spiritual or something bodily?[143] If something spiritual, he wonders if this could be that which is called heaven in Genesis 1:1 — that is, the first thing created. Thus, when the text states, "God said, 'Let light be made'; and light was made" (Gen 1:3), this is God calling heaven back to himself and perfecting his utterance. Why, Augustine asks, does the text state that "*God made*" heaven and earth and not "*In the beginning God said let heaven and earth be made*?" Further, in the creation

---

[140]Ibid.
[141]Ibid., 1.2.6 (TWSA I/13:170). Italics in original translation.
[142]Ibid. Italics in original translation.
[143]Ibid., 1.3.7 (TWSA I/13:170).

of light, why does Genesis indicate that "*God said, 'Let light be made'*"?[144] For Augustine it was possible that the expression *heaven and earth* needed to be understood in a universal way – that is, that it indicates everything created that was sensible and intelligible. After all was created there follows the explanation of how God made it "when it says of things one by one, *God said.*" Thus, God said – and made – everything through his Word.

Augustine's conclusion to the discussion about the "God made" and "God said" distinction in Genesis 1 involves a further distinction between Jesus as the Son of God, Jesus as Word, and Jesus as the "beginning" of Genesis 1:1. He has already indicated that the "heaven and earth" was formless matter. Creation imitates God the Word, "that is the Son of God who always adheres to the Father in complete likeness and equality of being, by which he and the Father are one [Jn 10:30]."[145] He appears to be stressing the connection of creation to himself. Just as the Son of God is eternally united to the Father, creation must be united to the Word. But if creation remains formless, it does not imitate the Word and turns away from the Creator. That is why Augustine sees the second person of the Trinity, the Son, in Genesis 1:1. "In the beginning *God made* . . . suggests the Word as the source of creation in its initial creation, its 'formless imperfection.'"[146] But when the text states "*God said, let it be made*," it is the Son who is being alluded to. His presence at the beginning means that he is the source of creation as it comes into being. But his being the Word indicates his conferring perfection on creation. The Word calls creation back to himself in order to complete it, "so that it may be given form by adhering to the creator."[147]

We have already seen the centrality of Christ for Augustine – "Moses," he says of Genesis, "wrote of him."[148] This centrality is reflected in his

---

[144]Ibid., 1.3.8 (TWSA I/13:171). Italics in original translation.
[145]Ibid., 1.49 (TWSA I/13:171).
[146]Ibid. Italics in original translation.
[147]Ibid.
[148]Augustine, *Confessions* 11.2.4 (TWSA I/1:225).

consideration of how God made all things: "By your Word you made them."[149] We should not miss the importance of the "Word" here. Augustine has already connected the Word with the beginning, creation, and Jesus. This is not a spoken word as in Matthew 3:17, when a voice came from a cloud saying, "This is my beloved Son." Utterances come and go; they have a beginning and an end and are heard through the movement of a created thing. The eternal Word of God is not like that – it is above us and abides forever.[150] Since the spoken word is heard because of the movement of created things, God clearly did not speak aloud in order to create. This would imply the existence of something prior to God since audible words require something to exist to be heard. Augustine explains,

> It seems, then, that if you made use of audible, evanescent words to say that heaven and earth should come to be, and that was how you made heaven and earth, there must have been some material thing in existence before you made heaven and earth, so that such a voice might use the creature's temporal movements to make itself audible in time. But no material thing did exist before heaven and earth.[151]

So, if it was not an audible word that was spoken at creation, what word was spoken to bring things into existence? "You are evidently inviting us to understand that the word in question is that Word who is God, God with you who are God, he is uttered *eternally*, and through him are *eternally* uttered all things."[152] Note here the connection between God, his Word, and the eternal utterance, along with the distinction Augustine has already made between eternity and time. The distinction is front and center – this is not a string of utterances where "one thing was said, and then, when that was finished, another thing, so that everything could be mentioned in succession."[153] Rather, this

---

[149]Ibid., 11.5.7 (TWSA I/1:227).
[150]Ibid., 11.6.8 (TWSA I/1:227-28).
[151]Ibid., 11.6.8 (TWSA I/1:228).
[152]Ibid., 11.7.9 (TWSA I/1:228). Italics mine.
[153]Ibid., 11.7.9 (TWSA I/1:228).

is a simultaneous utterance, "in one eternal speaking." This is necessary because if time and change were allowed to enter into the eternal speaking, there would be no eternity. The Word is coeternal with God. In him there is no cessation or succession; "all is truly immortal and eternal," even though the things he creates do not all come to be simultaneously, nor are they eternal.

Again we see the connection Augustine is making between the beginning, creation, and Jesus. All that comes to be and all that ceases to be does so in that eternal reason, "where nothing begins or comes to an end."[154] The eternal reason is explicitly identified as God's Word, who is "the beginning." This eternality represents stability for Augustine. It is in this beginning that God made heaven and earth. "You made them in your Word, your Power, your Wisdom, your Truth, wonderfully speaking in a wondrous way creating."[155]

The emphasis on the eternality of God, the Word, and the place of "the beginning" should be apparent here. Augustine is aware of the questions that this raises regarding temporality. By emphasizing the eternality of the Word and the "utterances" of the Word, there are clear implications for the issue of time and creation. Temporality is taken out of the equation in "the beginning" because these utterances are not audible. An utterance requires time and creation to be heard. But God did not "speak" this way through his Word. The "utterance" was eternal.

RECOMMENDED READING

Blowers, Paul. *Drama of the Divine Economy: Creator and Creation in Early Christian Theology and Piety.* Oxford Early Christian Studies. Oxford: Oxford University Press, 2012.

Christian, William A. "Augustine and the Creation of the World." *Harvard Theological Review* 46, no. 1 (1953): 1-25.

---

[154]Ibid., 11.8.10 (TWSA I/1:229).
[155]Ibid., 11.9.11 (TWSA I/1:229).

8

ON BEING *like* MOSES

The title of this final chapter may seem odd in light of the main topic of this book. Perhaps that is a signal of how far removed we are from ancient readings of the Bible that emphasize the theological and are not controlled by history. One would be very hard pressed to find in the earliest interpreters of sacred Scripture an approach that is intent only on finding direct one-to-one correspondence with strictly historical occurrences. Instead, one finds Scripture interpreted in a manner that emphasizes a call to a deeper spiritual life wherein the salvation of humankind and the ultimate goal of seeing God (contemplation) are overarching. Thus, in this final chapter I will explain how Basil uses Moses, the author of the creation narratives, as exemplar in that call.

To many of the church fathers, Moses was a progressive symbol for the spiritual life of the Christian, a type of every believer seeking to encounter God. It is not merely incidental, therefore, that Basil of Caesarea begins his entire homily series on the Hexaemeron with a section on the importance of the text's author. In doing so, he also shows the importance of the nature of the text and the required disposition of his hearers.[1] The author of the words "In the beginning God created the

---

[1] J. C. M. Van Winden, "'An Appropriate Beginning': The Opening Passage of Saint Basil's *In Hexaemeron*," in *Arche: A Collection of Patristic Studies by J. C. M. Van Winden*, ed. J. Den Boeft and D. T. Runia (Leiden: Brill, 1997), 116.

heavens and the earth" is praised for offering "an appropriate beginning" to the narrative for one who wants to speak about the creation of visible things.[2] It is appropriate because the orderly arrangement of visible things is handed down in the narration (*tou logou*) as having its origin in God, not happening spontaneously, as some have claimed.[3] The text, *tou logou*, is that which is handed down and constitutes the ground for the investigation.

But, according to Basil, his hearers are not mere listeners. They must be well prepared to hear about "such stupendous wonders"; otherwise, a devotion to the passions of the flesh will cloud investigation of a worthy concept of God.[4] After all, Basil later states, the investigation has the end goal of salvation of the hearers.[5] Since the text is the ground for investigating a worthy concept of God, it is important for Basil to establish the author's credentials. This is why Basil asks his hears to "consider who is speaking to us" before looking specifically at the text.[6]

BE LIKE MOSES

In Basil's insistence that his hearers "consider who is speaking to us," we should not miss the connection to the rhetorical training that was discussed in chapter three. Scholars have identified the hexaemeral sermons of Basil as fine examples of Christian rhetoric.[7] We have already learned that classical rhetoric was criticized by philosophers because they saw in it no real purpose except the glorification of the

---

[2]Basil, *Hexaemeron* 1.1 (FC 46:3). All quotations from *Hexaemeron* are from Saint Basil, *Exegetic Homilies*, trans. Agnes Clare Way, FC 46 (Washington, DC: Catholic University of America Press, 1963).

[3]See chapter five.

[4]Basil, *Hexaemeron* 1.1 (FC 46:3).

[5]Ibid. (FC 46:5).

[6]Ibid. (FC 46:3).

[7]J. M. Campbell, *The Influence of the Second Sophistic on the Style of the Sermons of St. Basil the Great*, Patristic Studies 2 (Washington, DC: Catholic University of America Press, 1922); and Hughes Oliphant Old, *The Reading and Preaching of the Scriptures in the Worship of the Christian Church*, vol. 2, *The Patristic Age* (Grand Rapids: Eerdmans, 1998).

speaker.[8] Orators in the ancient world received much praise and adoration because of their public acclaim. Basil is aware of this and wants to avoid the negative connection, so, at the very beginning of this series of homilies, he makes rhetoric "the servant of the Word," freeing it "from its long slavery to egotism."[9] This is why the text is his authority and why Moses is underscored as its author. He is authoritative as the author of the creation account because he spent forty years in Ethiopia contemplating creation.[10] By the time he was eighty years old, he "saw God as far as was possible for man to see him."[11] With Moses, God spoke "face to face – clearly, not in riddles."[12] The significance of Moses writing this creation account is highlighted because he saw God.

God revealed these things to Moses, so we are to accept them as from God himself. The congregation is exhorted to "hear . . . the words of truth expressed not in persuasive language of human wisdom but in the teachings of the Spirit, whose end is not praise from those hearing, but the salvation of those taught."[13] When we understand the context of rhetoric summarized above, it is difficult to miss this as a reference to it. There is, further, a clear allusion to Paul's words in 1 Corinthians 2:4: "My speech and my proclamation were not with plausible words of wisdom, but with a demonstration of the Spirit and of power." He sees himself in the same tradition as Paul in criticizing the self-seeking rhetor in search of praise.[14]

The authority of Moses, and thus the authority of Scripture, is grounded in the "teachings of the Spirit." Thus, the work of the Spirit through the text is necessary for hearers to understand what Moses wrote and, indeed, to be like Moses. The participatory nature of this

---

[8]Old, *Reading and Preaching of the Scriptures*, 2:46. See chapter three of the present work under "Ancient Greek Education – Philosophy and Rhetoric."

[9]Ibid.

[10]Cf. Ex 2:15: "But Moses fled from Pharaoh. He settled in the land of Midian."

[11]Basil, *Hexaemeron* 1.1 (FC 46:4).

[12]Ibid., citing Num 12:6-8.

[13]Ibid. (FC 46:4-5).

[14]Old, *Reading and Preaching of the Scriptures*, 2:47.

inquiry should be noted: the hearer is exhorted to prepare for these teachings of the Spirit through the words of Moses and to seek the same experience as Moses.

The exhortation to participation is made even clearer in the opening remarks of Basil's sixth homily on the Hexaemeron. There he compares the hearers of his homilies to the watchers of an athletic competition, and he encourages both to be more than mere observers. He invites his listeners to join with him in the "contemplation [*theōria*] of the wonders proposed" as a "fellow combatant," rather than a "judge."[15] The distinction between Basil's Christian rhetoric and the problems of classical rhetoric is also present here. The examination of the structure of the world and contemplation of the wonders proposed, he claims, is not from the wisdom of the world, but from Moses. God taught Moses "face to face – clearly, not in riddles." Just as Moses trained his mind to meet God, so Basil's hearers need to train their minds "for the consideration of what we propose." It is not in the same manner of "those who are fond of great shows and wonders."[16] What is clear is that the starting point for exploration and contemplation is what Moses wrote in the creation narratives themselves.[17] Be like Moses.

WHY WE AREN'T LIKE MOSES

Basil's call for his listeners to be well prepared to hear Scripture and his related warning against devotion to the passions of the flesh that cloud investigation of a worthy concept of God are anchored in his anthropology and the salvation of humanity, which is the goal of investigating a worthy concept of God. Basil does not direct much of his attention to the creation of humanity in his homilies on the Hexaemeron. Near the

---

[15]Basil, *Hexaemeron* 6.1 (FC 46:83).
[16]Ibid.
[17]Paul M. Blowers, "Entering 'This Sublime and Blessed Amphitheatre': Contemplation of Nature and Interpretation of the Bible in the Patristic Period," in *Nature and Scripture in the Abrahamic Religions*, vol. 1, *Up to 1700*, ed. Scott Mandelbrote and Jitse M. van der Meer, Brill's Series in Church History (Leiden: Brill, 2008), 147-48.

end of his ninth homily he explains that many listeners in his audience are "clamoring out" for explanations about the creation of humanity.[18] But the comment is conspicuous because in the previous eight homilies he had not broached the topic. Even here in Homily 9 his treatment of it focuses on the phrases "let us make mankind" and "in our image" as indications of the second person of the Trinity rather than on the creation of humanity proper. As he nears the end of the homilies he appears to sense a time crunch and promises to take up the topic later.[19]

The manuscript tradition that transmits the homilies on the Hexaemeron does not preserve any homilies beyond the ninth, but two more exist, preserved together, that many attribute to Basil.[20] It is in these two homilies where Basil apparently fulfills his promise to address the topic of the creation of humanity. While there is some scholarly debate about Basil's authorship of these two homilies, there is considerable agreement that they at least "reflect the mind of Basil."[21] Whether the homilies were the work of a stenographer who took notes from Basil[22] or the work of Basil himself, I will proceed on the assumption that these are at least Basilean in thought.

Indeed, the homilies go very well together as a whole.[23] At the beginning of Discourse 1, which some refer to as Homily 10, Basil makes reference to the promise he made in Homily 9: "I have come

---

[18]Basil, *Hexaemeron* 9.6 (FC 46:146-47).

[19]Ibid. (FC 46:149).

[20]Both homilies can be found in Nonna Verna Harrison, *St. Basil the Great on the Human Condition*, PPS 30 (Crestwood, NY: St. Vladimir's Seminary Press, 2005). In this English translation, the homilies are titled "First Homily. On the Origin of Humanity, Discourse 1: On That Which Is According to the Image," and "Second Homily. On the Origin of Humanity, Discourse 2: On the Human Being." Some refer to these homilies as Homilies 10 and 11. Since I am employing the Harrison translation here, I will retain her titles.

[21]Philip Rousseau, "Human Nature and Its Material Setting in Basil of Caesarea's Sermons on the Creation," *Heythrop Journal* 49 (2008): 222.

[22]This is the position of Maximos Aghiorgoussis, "Applications of the Theme 'Eikon Theou' (Image of God) According to Saint Basil the Great," *Greek Orthodox Theological Review* 21 (1976): 272.

[23]Rousseau, "Human Nature," 222, citing Alexis Smets and Michel Van Esbroeck, *Basile de Césarée, Sur l'origine de l'homme: Hom. x et xi de l' 'Hexaéméron' [par] Basile de Césarée*, introduction, text, French translation, and notes by Alexis Smets and Michel Van Esbroeck, *Sources chrétiennes*, 160 (Paris: Éditions du Cerf, 1970).

to make full payment of an old debt whose repayment I have postponed."[24] His delay, he says, was because of an illness, which may explain why these two homilies were transmitted separately from the previous nine.[25]

Basil takes great care in these homilies to emphasize the dignity of humanity. He exhorts his listeners to understand who they are because of the words in Genesis 1:26, "Let us make the human being." The human being was the only creation of God that was created after delib-eration—God did not say "Let there be a human being"; rather, there was counsel in the Godhead.[26] This unique event in creation gives hu-manity a certain dignity that Basil wants his hearers to understand: "Learn well your own dignity."[27]

Humanity, Basil continues, was created according to the image of God.[28] He counsels his hearers to expel any conception of God that is unworthy of him. "According to the image of God" is not to be under-stood as meaning that God is like us. Basil states:

> Do not enclose God in bodily concepts, nor circumscribe him according to your own mind. He is incomprehensible in greatness. Consider what a great thing is, and add to the greatness more than you have conceived, and to the more add more, and be persuaded that your thought does not reach boundless things. Do not conceive a shape; God is understood from his power, from the simplicity of his nature, not greatness in size. He is everywhere and surpasses all; and he is intangible, invisible, who indeed escapes your grasp. He is not circumscribed by size, nor encompassed by a shape, nor measured by power, nor enclosed by time, nor bounded by limits. Nothing is with God as it is with us.[29]

Human bodies change and are corruptible, but God is changeless and incorruptible. If the image of God does not consist in bodily shape,

---

[24]Basil, "On the Origin of Humanity, Discourse 1," 1 (PPS 30:31).
[25]Rousseau, "Human Nature," 223.
[26]Basil, Discourse 1.3 (PPS 30:33).
[27]Ibid.
[28]Ibid., 1.5 (PPS 30:34).
[29]Ibid.

what does it mean to say that humanity was created according to the image of God?

To answer the question Basil points to the final phrase of Genesis 1:26: "And let them rule over the fish."[30] Being created according to the image is connected with a ruling principle, with the ability to reason. For Basil, being created according to the image of God refers to our "inner being."[31] He speaks of "two human beings" – the one who can be apprehended according to sense, and the other who is "hidden under the sense-perceptible, invisible, the inner human."[32] The distinction is important because the outer things are not the essence of humanity – "For I am what concerns the inner human being, the outer things are not me but mine."[33] Thus, the body is an instrument of the soul because the soul is the human being.

The ruling principle of reason indicated by "and let them rule over the fish" is properly directed at the passions of the body, which are not included in the image of God – "reason is master of the passions."[34] This was given to humanity's nature; indeed, "your nature has the divine voice inscribed in it."[35] The dignity conferred on humanity by creation according to the image prioritizes the soul over the flesh. "Why," Basil asks, "do you throw away your own dignity and become a slave to sin? For what reason do you make yourself a prisoner of the devil? You were appointed a ruler over creation, and you have renounced the nobility of your own creation."[36] The original created nature of humanity had the ability to rule the passions of the flesh, but this nobility was renounced by humanity.

Humanity was created not only according to the image but also "according to the likeness." "The plan," says Basil, "had two parts."[37] Here

---

[30]Ibid., 1.6 (PPS 30:35).
[31]Ibid., 1.7 (PPS 30:35).
[32]Ibid., 1.7 (PPS 30:36).
[33]Ibid.
[34]Ibid., 1.8 (PPS 30:36).
[35]Ibid.
[36]Ibid. (PPS 30:37).
[37]Ibid., 1.15 (PPS 30:43).

we get further explanation on humanity's self-renunciation of nobility. Image and likeness are not the same thing. We have the image by our creation, but we "build" the likeness by free choice.[38] Humanity is conformed to that which is according to the likeness of God by free choice – "the power exists in us but we bring it about by our activity."[39] The power is afforded to humanity because of its creation according to the image. In fact, the free choice humanity exercises in becoming like God is the Christian life: "For I have that which is according to the image in being a rational being, but I become according to the likeness in becoming Christian."[40]

Becoming like God is about reining in passions away from selfish ends and directing them toward others. For example, forgiving one's enemy makes one like God, as does acting toward a sinner the way God does. In fact, becoming like God is synonymous with putting on Christ. "Take on yourself 'a heart of compassion, kindness,' that you may put on Christ. For through those things by which you undertake sympathy you put on Christ and drawing near to him is drawing near to God."[41] Thus, for Basil the deliberation in the Godhead to create humanity according to the image and likeness makes "the creation story . . . an education in human life."[42] The image is given, but the likeness is left incomplete so that humanity may complete it themselves. Yet the power of being created according to the image enables humanity to do so. This is precisely how Basil defines Christianity – "likeness to God as far as is possible for human nature."[43]

At the end of Discourse 1 Basil comes back to the scriptural injunction for humanity to rule over fish and beasts, equating that again with ruling the irrational passions.[44] Humans, he explains, have many

---

[38]Ibid., 1.16 (PPS 30:43).
[39]Ibid. (PPS 30:44).
[40]Ibid.
[41]Ibid., 1.17 (PPS 30:44).
[42]Ibid.
[43]Ibid. (PPS 30:45).
[44]Ibid., 1.19 (PPS 30:46-47).

beasts – anger, deceit, hypocrisy, greed, and lust. The power to rule means nothing if it is merely external. "Have you truly become ruler of beasts if you rule those outside but leave those within ungoverned?"[45] The creation story is an education in human life because "the rule we have been given over the animals trains us to rule the things belonging to ourselves."[46] There is more to the narrative of creation than the surface explanation given, just as there is more to the human being than the outer, external flesh.

Discourse 1 emphasized the dignity of human nature because of its creation according to the image of God. But in Discourse 2 Basil seeks to obviate an understanding of the flesh that may degrade it in light of his words in Discourse 1. Basil admits that his reading of Scripture gives a picture of humanity that is very pessimistic. Humanity is perishable and subject to the passions, enduring evils throughout life – all things he identified in Discourse 1. The problem is resolved in Genesis 2:7, which tells us that God took the "dust from the earth" and "molded the human being."[47] Basil says he learned that the human is both nothing and great. Humanity in itself is worthy of nothing, but it has been bestowed with great honor. The greatness is bestowed, as we learned from Discourse 1, because of the deliberation of the Godhead in the creation of humanity. But the greatness is bestowed also because "the Lord God took" – the human body is molded by God's own hands.

Here Basil makes a distinction between being "made" according to the image and God being said to "mold" the body. It is really another way of speaking about the inner and the external human being. When Psalm 119:73 states, "Your hands made me and molded me," the psalmist is showing the distinction – the inner being (the soul) was made and the outer being (the flesh) was molded.[48] Even though the distinction

---

[45]Ibid. (PPS 30:47).
[46]Ibid.
[47]Basil, Discourse 2.2 (PPS 30:49).
[48]Ibid., 2.3 (PPS 30:50).

is important, Basil emphasizes the honor that was bestowed on humanity in both:

> Ponder how you were molded. Consider the workshop of nature. The hand that received you is God's. May what is molded by God not be defiled by evil, not be altered by sin; may you not fall from the hand of God. Glorify your creator. For you came to be for the sake of no other thing except that you be an instrument fit for the glory of God. And for you this whole world is as it were a book that proclaims the glory of God, announcing through itself the hidden and invisible greatness of God to you who have a mind for the apprehension of truth. So be mindful of all the things that have been said.[49]

There is a clear connection between the molding of humanity and the making. Both give humanity honor. But the honor of the molding is renounced when the body is defiled by evil and sin. In this context, evil and sin should be understood from the perspective of Basil's claim that the greatness of God is available to those who have "a mind for the apprehension of truth" even though the greatness of God is "hidden and invisible." In other words, something can get in the way of humanity's apprehension of truth.

In order to explain what gets in the way of humanity's apprehension of truth, Basil directs his attention to Genesis 1:28: "And God blessed them and said, Grow and multiply and fill the earth."[50] The growth indicated here is of two kinds: growth of the body and growth of the soul. Growth of the body is natural; it is simply growth in physical stature. But growth of the soul is "progress to perfection through things learned."[51] We start to see here the connection Basil is trying to make with the aforementioned education. Irrational animals grow with regard to the body, but the inner human being grows according to the progress of growth into God. This is what Paul meant when he was "stretching out to the things before, forgetting the things behind [cf.

---

[49]Ibid., 2.4 (PPS 30:51).
[50]Ibid., 2.5 (PPS 30:51), as cited by Basil.
[51]Ibid.

Phil 3:13]."[52] Paul was, and we should be, reaching toward "truly existing things," and this entails growth in the inner human being in progress toward God.[53]

Multiply and *fill* in Genesis 1:28 also have meanings that point beyond the narrative. Thus, *multiply* refers to the proclamation of the gospel of salvation – multiply "those engendered according to the gospel."[54] *Fill* means to "fill the flesh" that has been given with good works. "Let the eye be filled with seeing duties. Let the hand be filled with good works. May the feet stand ready to visit the sick, journeying to fitting things. Let every usage of our limbs be filled with actions according to the commandments."[55]

Basil pauses to make an important qualifying statement that has implications for the interpretation of Scripture. He says, in reference to Genesis 1:28, that the words are common to irrational animals. That is, one can read the text and understand it as referring to animals. But "they have a specific meaning when we apply them to that which is according to the image and that with which we have been honored."[56] Thus, the anthropology so far preached by Basil is important context for understanding a more specific meaning, and this can be extended to how we are to read Scripture. Growing physically and growing spiritually are distinct yet related.

He continues in this vein by citing Genesis 1:29: "Behold, I have given you every tree which has fruit in itself; it will be to you for food."[57] As if to predict a reaction of his audience, Basil warns about skipping over this verse, because "all things are prescriptive."[58] The provision of fruit as food was given while humanity was still reckoned to be worthy of paradise. Basil says that the legislation of fishes, animals, reptiles, and quadrupeds

---

[52]Ibid. (PPS 30:51-52).
[53]Ibid. (PPS 30:52).
[54]Ibid.
[55]Ibid.
[56]Ibid.
[57]Ibid., 6.1 (PPS 30:52).
[58]Ibid.

for food was not originally given – only fruit. Every created being was sustained by fruit, and it was only later that concession was made for meat.

For Basil, there is a "mystery" in "what is hidden here."[59] Humanity "changed his habits and went outside the limit given him." Only after that did God grant them enjoyment of all foods. So there was some original condition from which humanity, by free choice, turned. This is what holds humanity back from being like Moses. The choice has proven to be a bad one, but restoration (salvation) is available. For Basil, this restoration

> after the present age will be such as was the first creation. And the human being will come again to his original condition, rejecting evil, this life's many troubles, the soul's enslavement involving life's concerns; putting aside all these things, he will return to that life in paradise un-enslaved to the passions of the flesh, free, intimate with God, with the same way of life as the angels.[60]

In this description it is difficult to miss the connection with Moses, who is a type of every believer seeking to encounter God. Because of the free choice to go outside the limits proper to their origin, humanity cannot be like Moses. The restoration of humanity entails a return to life in paradise, which means a life "unenslaved to the passions of the flesh, free, intimate with God." Be like Moses.

The variety of diet to which Basil refers here is really the introduction of sin. The mystery that is hidden is that humanity has fallen away "from the true delight that was in paradise. As such "we invented adulterated delicacies for ourselves."[61] The tree of life that bore the fruit of the first legislation of food is no longer in our sight; we have rejected that beauty and in its place have been given "cooks and bakers, and various pastries and aromas, and such things console us in our banishment from there."[62] There is something better for which humanity longs.

---

[59]Ibid. (PPS 30:53).
[60]Ibid., 7.7 (PPS 30:54).
[61]Ibid.
[62]Ibid.

At the end of Discourse 2, Basil again picks up the "molding" of humanity by God. Animals like sheep were created with heads inclining downward because they find their happiness in filling their stomachs. The human being does not look to the stomach but rather has head lifted high "toward things above, that he may look up to what is akin to him."[63] Basil explains that it is actually "against nature" to focus on earthly things; we should look instead to heavenly things where Christ is. Being molded is thus a lesson about the purpose for which humans were created. "You were born that you might see God, not that your life might be dragged down to earth, not that you might have the pleasure of beasts, but that you might achieve heavenly citizenship."[64] In the context, achieving heavenly citizenship is equated with the vision and contemplation of God.

Basil does not understand the narrative of the creation of humanity as one that details the method that God used to create. Using Moses as the example of the height to which humanity can ascend, he explains that humanity's own choice has obstructed its ability to be like Moses. Thus, there is something more going on here than a literal description of creation. We get a window into Basil's practice of a deeper reading of the text. He is not explicit about his methodology in this reading, but assumes its viability. In his homilies on the Hexaemeron and "Discourses on the Origin of Humanity," Basil does not really explain why he approaches the text in this manner. He does, however, give us a window into his approach in a few key places with his distinction between *theologia* and *oikonomia*.

THEOLOGIA AND OIKONOMIA

In Homily 9 Basil cites Genesis 1:26: "And God said, 'Let us make mankind.'" In explaining the phrase, he makes a curious comment in reference to Jewish understandings of this passage:

---

[63]Ibid., 2.15 (PPS 30:61).
[64]Ibid.

Where, I pray, is the Jew, who in times past, when the light of theology
[*theologias*] was shining as through windows, and the Second Person was
being indicated mystically [*mystikōs*], but not yet clearly revealed, fought
against the truth and said that God was speaking to Himself?[65]

A Jewish understanding of Genesis 1:26, according to Basil, is that God
was speaking to himself with the statement "Let us make mankind."
Basil thinks this is utter nonsense. Craftsmen like carpenters, copper-
smiths, or shoemakers simply do not set out to perform their craft
when alone and say, "Let us make" a plow, or sword, or shoe. Rather,
the work is simply done in silence. "Truly, it is utter nonsense for
anyone to sit down and command and watch over himself, and imperi-
ously and vehemently urge himself on."[66] For Basil, it is clear that the
passage refers to the second person of the Trinity, albeit "as long as the
one to be taught had not yet appeared, the preaching of theology [*theologias*] was deeply hidden."[67] For Basil, however, the Son had been revealed and is not now deeply hidden.

Twice within the same paragraph in Homily 9 Basil uses a form of
the word *theologia*. Noteworthy here is that, even though the context is
one in which Basil has promised to speak of the origin of humanity, he
focuses on the Father and the Son. Further, it is not on the works or
actions of creating humanity but rather on the identity of the Father
and Son. Thus, he asks his listeners to consider what "in our image"
means. Since it is in reference to the divine, it cannot mean a bodily
shape but a special reference to the Godhead. Citing various New Testament passages that refer to Christ as the image of the Father, he
emphasizes that intimate connection.[68]

Immediately before the passage from Homily 9 on the Jewish understanding of Genesis 1:26, Basil discusses the difficulty of even knowing
oneself. The sense of sight observes only external objects, so it is not

---

[65]Basil, *Hexaemeron* 9.6 (FC 46:147).
[66]Ibid.
[67]Ibid. (FC 46:148).
[68]Ibid. Basil cites Heb 1:3; Col 1:15; Jn 10:30; 14:9.

helpful. Even the mind is "slow in recognition of its own defects."[69] At the beginning of Discourse 1 Basil makes a similar reference to the difficulty of knowing oneself and the possibility of the mind being deceived in that endeavor. But this is why we need "divinely inspired Scripture" – because "our mind does not see itself" other than by examining Scripture. He continues, "For the light reflected there becomes the cause of vision for each of us. Since we are without understanding, we do not scrutinize our own structure; we are ignorant of what we are and why we are."[70] Basil encourages his listeners to consider "the wonder that is in you" because from the construction of the human body we see the "great Fashioner."

"Let us make the human being according to our image and likeness." As we have seen, the hortatory subjunctive, "Let us make the human being," is used only in reference to the creation of humanity. There was no "deliberation" for any other creative act of God.[71] For Basil, this indicates a certain dignity of humanity and relates to "the wonder that is within you." Why does Scripture indicate it this way instead of just God "making"? It is so that humanity may learn that the Father created through the Son, and that the Son created through the Father's will. Further, Scripture indicates this deliberation so that the Trinity may be glorified. Humanity has been made a "common work" of the Godhead, which should be worshiped together.

Basil then makes another statement in Discourse 1 that is reminiscent of one in Homily 9: "See a history [*historian*] in the form [of the Biblical passage] and theology [*theologian*] in [its] meaning."[72] It is significant that the statement is made here, as it is in Homily 9, not in the context of explaining the creation of humanity proper but rather in the context of the Godhead: "It says, 'Let us make,' that you may recognize

---

[69]Basil, *Hexaemeron* 9.6 (FC 46:148).
[70]Basil, Discourse 1.1 (PPS 30:31).
[71]Ibid., 1.3 (PPS 30:32-33).
[72]Ibid., 1.4 (PPS 30:33).

Father and Son and Holy Spirit."[73] In fact, it draws the hearers away
from fixating on the "how" of creation and toward focusing on who is
creating – the *historia* points to God. Basil sees it as a call to worship the
Creator who creates as Trinity. This is a vital connection for Basil because "the prelude to our creation is true theology [*theologia*]."[74]

The use of *theologia* here and in Homily 9 should not be understood
as it is commonly used today. That is, today *theology* is a general term
meant to indicate study about God. Basil does have this in mind, but
there is more at work here than just a simple description like "the
study of God." He is using it in the context of a technical distinction
from something called *oikonomia*. *Theologia*, as it was understood by
fourth-century Christian writers like Basil, was used in a restrictive
sense and concerned the divine nature (who God is). In Basil, it is a
"mode of insight into the nature of God," which is connected to seeing
beyond material reality or the "material-sounding phraseology" of
some passages in Scripture.[75] *Oikonomia*, on the other hand, concerned
the divine plan or the actions of God (what God does).[76] Thus, Basil
refers to an *oikonomia* in creation as well as in the incarnation and its
relationship to redemption.[77] Understanding the distinction is important because it comes into play with how Basil interprets the creation narratives. In other words, by using the word *theologia* Basil
signals an approach to Scripture that moves beyond the *historia* or narrative.[78] Frances Young hints at this direction:

> Critical evaluation of scriptural terms is essential, and that critical evalu-
> ation is based on the fundamental distinction between *theologia* and *oiko-
> nomia*, between the transcendent God – in principle unknowable and yet
> revealed in so far as the divine will and human limits allowed – and the

---

[73]Ibid.

[74]Ibid. (PPS 30:34).

[75]Lewis Ayres, *Nicaea and Its Legacy* (Oxford: Oxford University Press, 2004), 220.

[76]Mark Sheridan, *Language for God in the Patristic Tradition: Wrestling with Biblical Anthropomorphism*
(Downers Grove, IL: IVP Academic, 2015), 21, 26.

[77]Ayres, *Nicaea*, 220.

[78]See chapter three.

God whose loving plan involved the divine self in relationship with a created order quite other than the divine Being. There is no possibility of "narrative" in *theologia*, but narrative constitutes *oikonomia*; one is in time, the other beyond time.[79]

The distinction between *oikonomia* and *theologia* is explicitly present in Basil's important work *Against Eunomius*. He strenuously objects to Eunomius's claim that "begotten" in reference to Christ means the same things as "made." The biblical writers have no intention of communicating something about Christ's existence as "the very substance of God the Word, who was in the beginning with God."[80] This is beyond time and, therefore, inaccessible to us. The intent was to communicate something about his existence, in time, as the incarnation of the Word,

who "emptied himself in the form of a slave," and "became similar in form to the body of our lowliness," and "was crucified for our weakness." Everyone who has paid even marginal attention of the Apostle's text recognizes that he does not teach us in the mode of theology [*theologia*], but hints at the reasons of the economy [*oikonomia*].[81]

Scriptural talk about the Father begetting the Son, therefore, is not a reference to a temporal event that is performed for the sake of salvation. The text is, rather, speaking "theologically" – that is, about who God is. Conversely, when Scripture speaks of Christ taking on the form of a slave and being crucified, it is speaking "economically," or about what God does.[82]

The distinction is also seen when comparing Basil's view of the Synoptic Gospels (Matthew, Mark, Luke) to the Gospel of John. The Synoptic Gospels all speak of the incarnate Word's work. Matthew discusses the "begetting according to the flesh," Mark refers to the "beginning of the

---

[79]Frances Young, *Biblical Exegesis and the Formation of Christian Culture* (Peabody, MA: Hendrickson, 2002), 143.

[80]Basil, *Against Eunomius* 2.3 (FC 122:133), citing Jn 1:1. All quotations of *Against Eunomius* are from St. Basil of Caesarea, *Against Eunomius*, trans. Mark Delcogliano and Andrew Radde-Gallwitz, FC 122 (Washington, DC: Catholic University of America Press, 2010).

[81]Ibid., citing Phil 2:7; 3:21; and 2 Cor 13:4.

[82]Delcogliano and Radde-Gallwitz, introduction to *Against Eunomius*, 52 (FC 122).

Gospel of Jesus Christ," and Luke expounds on "corporeal origins."[83] This is not the case, however, with the writer of the Gospel of John, who did not repeat these corporeal origins but "needed to raise his mind above every sensory thing and time (which is concomitant to such things)."[84] Because the author of John's Gospel raised his mind above sensory things, Basil's opinion of the Gospel of John and its author surpasses that of the Synoptics. He "apprehended the beginning itself and left behind all corporeal and temporal notions as lower than his theology," and "he surpasses the preaching of the preceding evangelists on account of the nobility of his knowledge."[85] John's prologue does not speak of Mary as the mother of Christ, nor of times mentioned by other Gospel writers. Instead John speaks of *theology* – of the eternality of the Son, his begetting without passion, his sameness in being with the Father, and the majesty of his nature. The language here is similar to Basil's description of Moses.

There is one more place in Basil's work on the Hexaemeron where he draws on this distinction between *theologia* and *oikonomia* – Homily 6. In fact, with the context of the distinction covered above, we can use this homily as a way of understanding how Basil is a good representative of the general attitude of church fathers toward Genesis 1.

We have discussed Homily 6, on the creation of the lights of the heavens, fairly extensively. Here we will look at it from the angle of the distinction between *theologia* and *oikonomia* and how that might contribute to Basil's reading, or interpretation, of the creation account. As we have seen, Basil invites his hearers to join him in the "contemplation [*theōria*] of the wonders proposed."[86] The examination of the structure of the world is from Moses, the one whom God taught when he spoke to him "face to face – clearly, not in riddles." The mind, says Basil, needs to be "trained for the consideration of what we propose."[87]

---

[83]Basil, *Against Eunomius* 2.15 (FC 122:150).
[84]Ibid.
[85]Ibid.
[86]Basil, *Hexaemeron* 6.1 (FC 46:83).
[87]Ibid.

Basil expects his listeners to recall his first homily, which he had delivered only two days before, in which he offered a description of Moses that is very similar to the one he employed to the author the Gospel of John, who "apprehended the beginning itself."[88] The author of the Gospel of John was concerned with *theologia*, and Basil's description of Moses as the author of the Hexaemeron hints at this concern as well. Thus, the true starting point for exploration and contemplation is actually the creation narratives themselves, about which God speaks to Moses "mouth to mouth: and plainly, and not by riddles."[89]

The hint of a concern for *theologia* is confirmed in Homily 6. After reminding his listeners of the author of the Hexaemeron, Basil then explains a bit more about the concern for *theologia*. Earlier in Homily 4, he refers to the *improper* use of creation as a "theatre, flourishing with impure sights," which moves participants to "unseemly behavior."[90] The pleasurable, material aspects of the world do not sustain a pure life. In Homily 6, Basil takes the opposite approach and explains the *proper* use of creation:

> If, at any time in the clear cool air of the night, while gazing intently at the indescribable beauty of the stars, you conceived an idea of the Creator of the universe – who He is who has dotted the heavens with such flowers, and why the usefulness is greater than the pleasure in visible things or again, if at times you observed with sober reflection the wonders of the day and through visible things you inferred the invisible Creator, you come as a prepared listener and one worthy to fill up this august and blessed assembly.[91]

There is a proper stance or attitude ("preparation") with which listeners and readers must approach the creation account, says Basil. The visible things of creation were not intended to be an end in themselves but to "infer the invisible Creator."

---

[88]Basil, *Against Eunomius* 2.15 (FC 122:150).
[89]Basil, *Hexaemeron* 1.1 (FC 46:4). Cf. Blowers, "Entering," 1:147-48.
[90]Basil, *Hexaemeron* 4.1 (FC 46:55).
[91]Ibid., 6.1 (FC 46:84).

Basil urges his audience, as fellow participants in this great universe of wonder, to take his hand and allow him to guide them, "as strangers, to the hidden wonders of this great city," which is actually "our ancient home" from whence we were driven by the devil.[92] Humanity is the creation of God "from the earth," but it is far above the animals in the ability to "be lifted up to the very heavens" by the "benefit of reason [*logou*]."[93] The animals were "without reason" (*alogōn*); humanity was the creation imbued with rationality. The English translation of *logou* as "reason" inclines us to think of it as having something to do with rationality. But, as we have seen, we should understand it as Basil does, in connection with the ability to rule that was given during creation according to the image of God. The contrast between humanity having the "benefit of reason" and animals being "without reason" should remind us of our dignity.

Knowledge of this, says Basil, allows for knowledge of ourselves and God and leads to worship of him as Creator of the beauty and wonder we see. But the beauty and wonder we see are not ends in themselves. Our problem is that we have considered them such. But, as Basil states,

> If transient things are thus, what will be the eternal? And if visible things are so beautiful, what will be the invisible? If the grandeur of the heavens transcends the measure of the human intellect, what mind will be able to explore the nature of the everlasting? If the sun, subject to destruction, is so beautiful, so great, so swift in its motion, presenting such orderly cycles, possessing a magnitude so commensurate with the universe that it does not exceed its due proportions to the universe; if by the beauty of its nature it is as conspicuous in creation as a radiant eye; if the contemplation of it is incapable of satisfying us, what will be the beauty of the Sun of justice [Mal 4:2]? If it is a loss to a blind man to be unable to look upon this, how great a loss is it to a sinner to be deprived of the true Light?[94]

---

[92]Ibid.
[93]Ibid.
[94]Ibid. (FC 46:84-85).

There is a beauty and wonder in creation that provokes awe. But Basil is insistent that this is nothing compared to that which transcends, orders, and inheres in that beauty. Missing this *real* beauty is akin to a blind man who is unable to look on the beauty of the heavens – thus, the *true* Light exists, but many miss it, and this indeed is a "great loss."

Clearly the "true Light" referred to by Basil is a move to look deeper in the narrative. He confirms this when he refers to the words of Genesis 1:14-15: "And God said, 'Let there be lights in the firmament of the heavens for the illumination of the earth, to separate day from night.'"[95] He does not immediately begin to explain what he means by contemplating the "Sun of justice" or the "true Light." Instead he sticks with the narrative of the text and explains the order of creation of heaven, earth, day, and night. The sun was not in existence yet, says Basil, so that no one could call it the first cause of light rather than God. But these explanations are not at the forefront of his agenda. Concerning the scriptural command of God, "Let there be light," Basil asks, "Who spoke and who made? Do you not notice in these words the double Person? Everywhere in history [*historia*] the teachings of theology are mystically interspersed [*theologias mystikōs symparespartai*]."[96]

The *theologia* that is "mystically interspersed" with *historia* is concerned not with the creation of the sun and the moon but rather with the "pure, clear, and immaterial light," for which they are a vehicle.[97] Scripture does attest to material light in the world, but Basil's concern is more *theological* in the sense we have been discussing: "But the true Light of the world is something else, and by participation in it holy men become the lights of the souls whom they have taught, drawing them out from the darkness of ignorance, so also now, having prepared this sun for that most bright light, the Creator of the universe has lighted it around the world."[98]

---

[95]Ibid., 6.2 (FC 46:85), as cited by Basil.
[96]Ibid.
[97]Ibid. (FC 46:86).
[98]Ibid.

RETURN TO PARADISE

As we have seen, writers on both the AiG and CMI websites have singled out Basil as a literalist in the same sense they understand the term. They have used Basil to speak for the biblical exegesis of creation science. This is a disservice not only to Basil but also to those who look to the church fathers for guidance in their own Christian walk and interpretation of Scripture.

Chapter three engaged directly with the assertion that Basil was a literalist. There, I concluded that Basil's various comments about allegory appeared out of character for the great Cappadocian and thus needed to be understood in context. In that light, Basil cannot be labeled as one who rejected all allegory in favor of the literalism espoused by creation science apologists. In the present chapter, I have gone beyond my own apologetic against that view and toward a more positive understanding of the main thrust of Basil's exegesis.

Basil's primary interest is not in the events *behind* the text, and his understanding of the creation of humanity shows this. I have suggested that Basil is asking his listeners to be like Moses. This is synonymous with the restoration of humanity returning to paradise, which means a life "unenslaved to the passions of the flesh, free, intimate with God."[99]

Returning to paradise is the restoration and goal of the human life, and this is what the creation story ultimately shows. Consoling ourselves with material things keeps us from paradise. This is why the story of creation is "an education in human life"[100] and why the story of creation should be read with an eye toward that end. The history *behind* the text does not concern Basil, but the *theologia* behind it does.

RECOMMENDED READING

Blowers, Paul M. "Entering 'This Sublime and Blessed Amphitheatre': Contemplation of Nature and Interpretation of the Bible in the Patristic Period." In

---

[99]Basil, Discourse 2.7 (PPS 30:54).
[100]Basil, Discourse 1.17 (PPS 30:44).

Nature and Scripture in the Abrahamic Religions. Vol. 1, *Up to 1700*, edited by Scott Mandelbrote and Jitse M. van der Meer, 147-74. Brill's Series in Church History. Leiden: Brill, 1978.

Harrison, Nonna Verna. *God's Many-Splendored Image: Theological Anthropology for Christian Formation.* Grand Rapids: Baker Academic, 2010.

Rousseau, Philip. *Basil of Caesarea.* Berkeley: University of California Press, 1994.

SELECTED BIBLIOGRAPHY

PRIMARY SOURCES

Augustine. *The City of God, Books 11–22*. Translated by William Babcock. TWSA I/7. Hyde Park, NY: New City Press, 2013.

———. *The Confessions*. Translated by Maria Boulding. TWSA I/1. Hyde Park, NY: New City Press, 1997.

———. *On Genesis*. Translated by Edmund Hill. TWSA I/13. Hyde Park, NY: New City Press, 2002.

Basil of Caesarea. *Exegetic Homilies*. Translated by Agnes Clare Way. FC 46. Washington, DC: Catholic University of America Press, 2003.

———. *On the Human Condition*. Translated by Nonna Verna Harrison. PPS 30. Crestwood, NY: St. Vladimir's Seminary Press, 2005.

Ephrem the Syrian. *Selected Prose Works: Commentary on Genesis; Commentary on Exodus; Homily on Our Lord; Letter to Publius*. Translated by Edward G. Mathews and Joseph P. Amar. FC 91. Washington, DC: Catholic University of America Press, 2004.

Origen of Alexandria. *On First Principles*. Translated by G. W. Butterworth. Glouster, MA: Peter Smith, 1973.

Theophilus of Antioch. *Ad Autolycum*. Translated by Robert M. Grant. OECT. Oxford: Clarendon Press, 1970.

SECONDARY SOURCES

Allert, Craig D. *A High View of Scripture? Biblical Authority and the Formation of the New Testament Canon*. Evangelical *Ressourcement* Series. Grand Rapids: Baker Academic, 2007.

Blowers, Paul M. "Doctrine of Creation." In *The Oxford Handbook of Early Christian Studies*, edited by Susan Ashbrook Harvey and David G. Hunter, 906-31. Oxford: Oxford University Press, 2008.

———. *Drama of the Divine Economy: Creator and Creation in Early Christian Theology and Piety*. Oxford Early Christian Studies. Oxford: Oxford University Press, 2012.

———. "Entering 'This Sublime and Blessed Amphitheatre': Contemplation of Nature and Interpretation of the Bible in the Patristic Period." In *Nature and Scripture in the Abrahamic Religions*. Vol. 1, *Up to 1700*, edited by Scott Mandelbrote and Jitse M. van der Meer, 147-74. Brill's Series in Church History. Leiden: Brill, 2008.

Bouteneff, Peter C. *Beginnings: Ancient Christian Readings of the Biblical Creation Narratives.* Grand Rapids: Baker Academic, 2008.

Christian, William A. "Augustine and the Creation of the World." *Harvard Theological Review* 46, no. 1 (1953): 1-25.

Copan, Paul, and William Lane Craig. *Creation Out of Nothing: A Biblical, Philosophical, and Scientific Exploration.* Grand Rapids: Baker Academic, 2004.

Furley, David. *The Greek Cosmologists.* Vol. 1, *The Formation of the Atomic Theory and Its Earliest Critics.* Cambridge: Cambridge University Press, 1987.

Hall, Christopher A. *Learning Theology with the Fathers.* Downers Grove, IL: InterVarsity Press, 2002.

——. *Reading Scripture with the Fathers.* Downers Grove, IL: InterVarsity Press, 1998.

Harrison, Nonna Verna. *God's Many-Splendored Image: Theological Anthropology for Christian Formation.* Grand Rapids: Baker Academic, 2010.

Lim, Richard. "The Politics of Interpretation in Basil of Caesarea's 'Hexaemeron.'" *Vigiliae Christianae* 44, no. 4 (1990): 351-70.

Litfin, Bryan. *Getting to Know the Church Fathers: An Evangelical Introduction.* Grand Rapids: Brazos Press, 2007.

Louth, Andrew. *Discerning the Mystery: An Essay on the Nature of Theology.* Oxford: Clarendon Press, 2003.

——. "The Six Days of Creation According to the Greek Fathers." In *Reading Genesis After Darwin,* edited by Stephen C. Barton and David Wilkinson, 39-55. Oxford: Oxford University Press, 2009.

May, Gerhard. *Creatio Ex Nihilo: The Doctrine of "Creation Out of Nothing" in Early Christian Thought.* Translated by A. S. Worrall. London: T&T Clark, 2004.

McGuckin, John A. "Patterns of Biblical Exegesis in the Cappadocian Fathers: Basil the Great, Gregory the Theologian, and Gregory of Nyssa." In *Orthodox and Wesleyan Scriptural Understanding and Practice,* edited by S. T. Kimbrough, 37-54. Crestwood, NY: St. Vladimir's Seminary Press, 2005.

Mook, James R. "The Church Fathers on Genesis, the Flood, and the Age of the Earth." In *Coming to Grips with Genesis: Biblical Authority and the Age of the Earth,* edited by Terry Mortenson and Thane H. Ury, 23-51. Green Forest, AZ: Master Books, 2008.

O'Keefe, John J. "'A Letter That Killeth': Toward a Reassessment of Antiochene Exegesis, or Diodore, Theodore, and Theodoret on the Psalms." *Journal of Early Christian Studies* 8, no. 1 (2000): 83-103.

O'Keefe, John J., and R. R. Reno. *Sanctified Vision: An Introduction to Early Christian Interpretation of the Bible.* Baltimore: Johns Hopkins University Press, 2005.

Orphanos, M. A. *Creation and Salvation According to St. Basil of Caesarea.* Athens: G. K. Parisianos, 1975.

Pelikan, Jaroslav. "The 'Spiritual Sense' of Scripture: The Exegetical Basis for St. Basil's Doctrine of the Holy Spirit." In *Basil of Caesarea: Christian, Humanist, Ascetic,* edited by Paul Jonathan Fedwick, 337-60. Toronto: Pontifical Institute of Medieval Studies, 1981.

Peters, Edward. "What Was God Doing Before He Created the Heavens and the Earth?" *Augustiniana* 34 (1984): 53-74.

Robbins, Frank Egleston. *The Hexaemeral Literature: A Study of the Greek and Latin Commentaries on Genesis.* Chicago: University of Chicago Press, 1912.

Sandwell, Isabella. "How to Teach Genesis 1.1-19: John Chrysostom and Basil of Caesarea on the Creation of the World." *Journal of Early Christian Studies* 19, no. 4 (2011): 539-64.

Trigg, Joseph W. *Biblical Interpretation.* Message of the Fathers of the Church 9. Wilmington, DE: Michael Glazier, 1988.

Watson, David. "An Early View of Genesis One." *Creation Research Society Quarterly* 27, no. 4 (1991): 138-39.

Wickes, Jeffrey. "Ephrem's Interpretation of Genesis." *St. Vladimir's Theological Quarterly* 52, no. 1 (2008): 45-65.

Wilken, Robert L. *The Spirit of Early Christian Thought: Seeking the Face of God.* New Haven, CT: Yale University Press, 2003.

Williams, D. H. *Evangelicals and Tradition: The Formative Influence of the Early Church.* Evangelical *Ressourcement* Series. Grand Rapids: Baker Academic, 2005.

———. *Retrieving the Tradition and Renewing Evangelicalism: A Primer for Suspicious Protestants.* Grand Rapids: Eerdmans, 1999.

———. "*Similis et Dissimilis*: Gauging Our Expectations of the Early Fathers." In *Ancient Faith for the Church's Future,* edited by Mark Husbands and Jeffrey P. Greenman, 69-89. Downers Grove, IL: IVP Academic, 2008.

Young, Frances M. *Biblical Exegesis and the Formation of Christian Culture.* Peabody, MA: Hendrickson, 2002.

———. "The Fourth Century Reaction Against Allegory." *Studia Patristica* 30 (1997): 120-25.

———. "The Rhetorical Schools and Their Influence on Patristic Exegesis." In *The Making of Orthodoxy: Essays in Honour of Henry Chadwick,* edited by Rowan Williams, 182-99. Cambridge: Cambridge University Press, 1989.

———. "Typology." In *Crossing the Boundaries: Essays in Biblical Interpretation in Honour of Michael D. Goulder,* edited by Stanley E. Porter, Paul Joyce, and David E. Norton, 29-48. Leiden: Brill, 1994.

AUTHOR INDEX

SUBJECT INDEX

SCRIPTURE INDEX

Finding the Textbook You Need